Praise for *It Takes a Genome*

"A compelling, witty, and reader-friendly explanation of how our genes, fashioned for living in the Stone Age, are not so well-suited to life in the Modern Age."

—Sean B. Carroll, author of *The Making of the Fittest* and *Remarkable Creatures*

"It's taken thirty years, but we finally have in Greg Gibson's *It Takes a Genome* what is truly a biologist's response to the single-gene focus of Richard Dawkin's early classic *The Selfish Gene*. And what a response it is! In Gibson's world, we see a genome as an integrated whole, making sense only when the constituent parts, the genes, are considered in their full genomic and environmental context. It is an engaging, fascinating, accessible, and ultimately deeply satisfying perspective that will enrich the way we all think about ourselves and how we got to be the way we are."

—David B. Goldstein, Professor of Molecular Genetics, Duke University

"Gibson has captured the delicate balance between the excitement of the genomic revolution and the frustration that so much is yet to be learned about the genomics of disease. This book is an ideal guide through the complexities of recent environmental change and how this non-genetic process has interacted with human genomic variation to produce today's landscape of important chronic diseases."

—Marc Feldman, Professor of Biology, Stanford University

"Gibson deftly synthesizes the new science linking genome variation and human health, debunking entrenched views about the causes and evolution of disease and arguing convincingly for a more comprehensive view. An important book and a great read."

—David P. Mindell, Dean of Science, California Academy of Sciences

"Geneticist Gibson is a natural teacher. He brings a welcome balance to his descriptions of the roles of genes, the environment, and chance in the major human diseases."

—Bruce Weir, Chair and Professor of Biostatistics, University of Washington

It Takes a Genome

It Takes a Genome

*How a Clash Between Our Genes and
Modern Life Is Making Us Sick*

Greg Gibson

Vice President, Publisher: Tim Moore
Associate Publisher and Director of Marketing: Amy Neidlinger
Senior Acquisitions Editor: Amanda Moran
Editorial Assistant: Myesha Graham
Development Editor: Russ Hall
Digital Marketing Manager: Julie Phifer
Publicity Manager: Laura Czaja
Assistant Marketing Manager: Megan Colvin
Cover Designer: Stauber Design
Operations Manager: Gina Kanouse
Managing Editor: Kristy Hart
Project Editor: Chelsey Marti
Copy Editor: Geneil Breeze
Proofreader: Debbie Williams
Senior Indexer: Cheryl Lenser
Compositor: Jake McFarland
Manufacturing Buyer: Dan Uhrig

© 2009 by Pearson Education, Inc.
Publishing as FT Press Science
Upper Saddle River, New Jersey 07458

FT Press Science offers excellent discounts on this book when ordered in quantity for bulk purchases or special sales. For more information, please contact U.S. Corporate and Government Sales, 1-800-382-3419, corpsales@pearsontechgroup.com. For sales outside the U.S., please contact International Sales at international@pearson.com.

Second Printing March 2009
ISBN-10: 0-13-713746-X
ISBN-13: 978-0-13-713746-6

Pearson Education LTD.
Pearson Education Australia PTY, Limited.
Pearson Education Singapore, Pte. Ltd.
Pearson Education North Asia, Ltd.
Pearson Education Canada, Ltd.
Pearson Educatión de Mexico, S.A. de C.V.
Pearson Education—Japan
Pearson Education Malaysia, Pte. Ltd.

Library of Congress Cataloging-in-Publication Data is on file.

For Diana

Contents

Preface

*How a genetic culture clash with
modern life is making us sick*

We humans, I am sure I have little need to convince you, are an extraordinary species. Whether you regard us as the pinnacle of Creation or the latest exemplar of the evolutionary process, our genes endow us with a certain uniqueness. We are capable of great athleticism, artistic genius, bravery, brilliance, creativity, and conscious reflection. No other species has our linguistic dexterity, or the reasoning skills that have led us in a few short millennia to dominate the planet. Yet for all this wonder, we are also astonishingly genetically vulnerable. More than half of us will die of a complex disease whose origins can be traced to genetic susceptibilities that place us at risk in the modern environment of our making.

The seeds of our discontent lie hidden in the human genome, uncovered now by a genetic culture clash with contemporary life. It turns out that organisms evolve not just to approach some optimum, but also to be buffered against the vagaries of circumstance. Take any species outside its comfort zone, and all of a sudden it gets a whole lot more vulnerable. In the last few hundred years, humans have created an environment defined by fast and sugary foods and bland immune exposure, while our mental world is shaped more by electronic energy than the soft sensibilities of the biosphere. Is it any wonder that diabetes, asthma, and depression are almost epidemic?

The mission of this book is to explain how our genes make us sick. Secondarily, it is to advance the thesis that they do so in large part because the genome is out of equilibrium, with itself and with the environment. If you like, our genes are "not in a happy place." So much has changed so quickly in human history, starting 10,000 generations ago with the origin of the species and accelerating 10 decades ago with the pace of industrialization, that many genetic processes are not quite right. The stress of modernity provides a little extra shove that pushes otherwise perfectly normal varieties of genes to the brink of malfunction. Like a bad casserole, some flavors just don't go together, certainly not with the ingredients they are being paired with. Flavors that throughout primate history have been perfectly innocuous now find themselves singled out as the bad guys, as the risk factors that contribute to obesity and inflamed bowels and kids who can't pay attention.

All I ask of you is to suspend some of the beliefs that you may have picked up from the media, or from fundamentalist Darwinians. It is convenient for journalists to write stories about genes for this or for that: genes for aggression and genes for altruism; good genes and selfish genes. But it is the way that genes work together inside cells that influences whether your nose is long or your girth is rotund. Every single gene comes in a variety of types, some common and some more rare, and just about every gene has multiple tasks and responsibilities. The key to understanding why they make us sick is to understand that, just like every one of us, they are just trying to do their best given the features they were endowed with in a complicated world. With the best intentions, sometimes things just don't work out—particularly when we're outside our comfort zone.

This book is written from the perspective of an empirical evolutionary quantitative geneticist. A what? This is actually a mainstream branch of biological research, one populated by several faculty members on most university campuses from New York City to Manhattan, Kansas, and that underlies an increasing volume of the genetics you read about. The "quantitative" part is a code word for concepts such as complex and diverse and statistical estimation. It is the genetics you never learn in high school but certainly should, because it is the genetics of everything we encounter on a daily basis. Height, color, degree of spirituality, and disease susceptibility all have tendencies to be transmitted from parent to child, and all are influenced by many genes interacting with the environment. The Human Genome Project has brought the study of these interactions to our fingertips, and much of what I have to say is about this new learning.

The "evolutionary" part is a nod to history, recognition that we are a product of our past. We can certainly study genetics with a singular focus on the here and now, possibly going back just a generation or two, but everything makes so much more sense when we see things in the context of millions, and even hundreds of millions, of years of life on Earth. Nary a serious practicing research geneticist does not embrace the facts and theory of evolution. The core theory is actually due to twentieth century luminaries such as R. A. Fisher, J. B. S. Haldane, Sewall Wright, Richard

Lewontin, and Motoo Kimura, who built genetics into the edifice founded by the grand old Victorian Charles Darwin. There are, of course, lively debates about the details of the process, but there is little disagreement with the realization that the inclusion of history helps us understand how things work. The past has shaped the present just as much in our genes as in our beings.

The "empirical" part places the emphasis on observation and experimentation. Theory plays a crucial role, but alone it can be horribly misleading. Certain aspects of theoretical genetics have garnered popular attention, most notably the more rabid strains of sociobiology that equate human behavior with the lekking of a bowerbird or the castes of a honeybee. Yet it is only by getting our hands dirty with real data can we sort the actual from the possible. Reams and reams of data are now at hand, generated by automated DNA sequencers that churn out millions of bits of information every second in high-tech genetic power plants that have popped up like mushrooms across our campuses. A scientific revolution is in progress, and this too is the subject of the book.

We will follow an orthodox strategy. The six core chapters deal with each of six major classes of human disease sequentially: cancer followed by diabetes, inflammatory and infectious diseases, and then two domains of psychology, depression and senility. These six chapters are sandwiched between an introduction that lays the conceptual foundation and a conclusion that offers some thoughts on human diversity more generally.

Chapter 1, "The Adolescent Genome," explains how genes are for the most part extraordinarily interactive and cooperative entities—"it takes a genome" to build most traits. We'll consider the notion of molecular existentialism, which is a way of saying that what a gene does is very much a function of whom it knows and whom it works with on a daily basis. Here one gene affects the hardness of your tooth enamel, there the growth of your cranial nerves, and elsewhere it helps to ensure that your liver is the right size. It does not make much sense to talk about a gene for eyes or a gene for altruism. Rather, we need to think of things in terms of variation. Every single gene comes in a variety of flavors, and whether you are taller or shorter, more or less prone to diabetes, or better at shooting a basketball than the next guy, is influenced by the flavors you get at dozens if not hundreds of genes. They work together in intricate networks that in a sense breathe with the environmental variation as well. Cystic fibrosis and muscular dystrophy aside, this leads to the idea that disease is most often not the result of a single bad gene, but instead is caused because the genome is not yet mature, and there is a mismatch between combinations of normal varieties of gene and unfavorable circumstances.

Chapter 2, "Breast Cancer's Broken Genes," concentrates on breast cancer. It is a bit of a strange place to begin our survey of the genetics of complex disease, because the way genes cause cancer is very different from the manner in which they influence the other diseases. For one thing, they really cause it. Once the genes inside a cell are broken beyond repair, there is nothing to be done other than to control the damage either by cutting out the tumor, or destroying it by cutting off the blood supply or actively killing the cells. Certainly a healthy spirit helps, but there is a shocking inevitability about cancer. For another, it is not the genes we get from our parents that matter, for the most part, but rather the things that happen to them during our lives. Genes break, just as faucets and gutters and washing machines and radiators break in the course of wear and tear. They come with a warranty in the form of an active toolkit for repairs, but when the repairmen themselves break down, trouble really sets in, most notably in breast and colon cancer. Cancers take multiple mutations to get started, and each of us inherits a set of mutations whose effects are veiled by the rest of the healthy genome, only to exert their effects as other mutations accumulate. Nor are there really cancer genes, in the sense that there are not genes whose role it is to promote cancer: What we call cancer genes are just normal, for the most part essential, genes that we need for normal growth and development, and when these are broken, they lead to cancer. Importantly too, it needs to be emphasized that much of the recent increase in cancer rates has been brought on by human activities—some that we don't want to do much about, such as growing older and maturing earlier, and some that we can control, such as avoiding smoking or sunbathing all day. In the end, cancer is as much about genes interacting with the modern environment as any other disease.

In Chapter 3, "Not So Thrifty Diabetes Genes," our attention turns to diabetes. One type of diabetes is due to immunological problems, but the predominant type is allied with obesity. We know now that obesity and type 2 diabetes are joined with cardiovascular disease and stroke in a spectrum of discomfort known as metabolic syndrome that affects almost a third of all westerners. These diseases all trace in part to the incapacity of our genes to cope with contemporary lifestyles. The title is a play on a popular hypothesis in evolutionary medicine that disease emerges as a result of conflict between selection in the past and the needs of the present. Supposedly, genes that helped us survive famine by rapidly storing fat now promote diabetes by slowly accumulating it in excess. However, the data really don't support this, and instead I develop the concept of disequilibrium between veiled variation that in the past had little or no effect on human physiology and the energy dense diet of our times. In fact, metabolic syndromes are the poster child for genetics in the era of genome biology, as the application of brute-force screening has revealed the identity of dozens of risk factors. Something called the common disease-common variant hypothesis seems to hold true. Diabetes really does trace, in part, to different flavors of genes that perhaps a quarter of us have, and it is the combinations of these common flavors with environmental factors that predicts disease—"predicts" in the sense of epidemiology, not necessarily an individual. Someone with every hedonistic tendency in the world who has even been dealt a sorry deck of genetic cards can nevertheless live long and prosper, while another poor soul who sticks to the Atkins Diet may find him- or herself overweight and encumbered with a failing heart in middle age.

Chapter 4, "Unhealthy Hygiene," turns to a discussion of inflammatory immune disorders. There's a litany of A's: asthma, atopy, allergies, and arthritis, in addition to the autoimmunity of inflammatory bowel disease, lupus, and multiple sclerosis. Here the story is one of shifting balances. Our immune systems are set up to protect us against three classes of enemies—viruses, bacteria, and parasites—while avoiding the perils of friendly fire. They operate in a constantly changing battlefield and deploy constantly against shifting artillery. Moreover, the nature of the threat has changed a couple of times in human history, first with the increased exposure to zoonotic disease, as we adopted pets and domesticated livestock; and more recently as we moved into crowded cities and/or sterile suburbs. As much as one-fifth of our genome is set up to help us cope with pathogens, operating in a complex web of interactions full of checks and balances. The challenge of our time is not to overreact to perceived threats without doing more harm than good, but the confusion in our genome, pushed to the edge of its buffering capacity, emerges in the form of inflammation and irritation.

A short Chapter 5, "Genetic AIDS," is reserved for the other side of immune malfunction, the Human Immunodeficiency Virus (HIV) and Acquired Immune Deficiency Syndrome (AIDS). This disease is a lightning rod for discussion of so many of the deep moral concerns of humanity, bridging notions of individual responsibility (for behaviors that often lead to infection) to the burden of care of the developed world to the developing world (while respecting local governmental autonomy). We will see that widespread variation in our genes determines who will be protected and who will be able to cope with chronic infection, and we will see that this variation is unevenly distributed among different ethnicities. Further, it seems that some of the protection can be attributed to past episodes of response to bygone plagues and epidemics.

The focus shifts to behavior in Chapter 6, "Generating Depression," which is all about psychological diseases. Profound sadness and anxiety may yet join diabetes as the defining discomforts of the twenty-first century. Alcoholism and other addictions of abuse blight the lives of those unfortunate enough to suffer the pain they cause, as do the health policies of societies that refuse to acknowledge that the capacity for self-control is not equally distributed. No one can seriously doubt any longer that genes contribute to psychological maladies, but, unfortunately, geneticists cannot yet see a coherent picture of their role. Perhaps thousands of rare mutations contribute, or perhaps here is where the misfit between our genomic heritage and contemporary life is at its starkest. Kids these days learn their way around an Xbox or a PlayStation before they can tell an oak from an elm, or a cirrus from a cumulus cloud wafting across the skies. Every aspect of our mental world, be it relationships, careers, or pastimes, is new, and we careen through life at a pace that we just did not evolve for. Surely the pressures on the nervous system are at least as great as those on our metabolism and immune systems, so here too disease emerges from the unveiling of normal variation under the stress of modernity.

The final core chapter, Chapter 7, "The Alzheimer's Generation," is appropriately placed at the end of our survey of genetic discomfort and deals with senility and growing old. This chapter recognizes that every adult becomes conscious of the dementias in their middle age when their thoughts turn to care of aging loved-ones. There's plenty of evolutionary theory about conflicts between genes that are good for us when we're young but turn against us as we age, or about genes not caring so much about keeping us alive after we've raised our offspring. However, some of us are predisposed genetically to decay mentally at an accelerated rate, and modern genomics tells this one last story of disequilibrium between our genetic past and contemporary lifestyle.

I need to emphasize that despite all this, I am not advocating a turning back of the clock or a return to some lost utopia that never was. To use a well-worn phrase, modernity "is what it is." It is unrealistic to expect people to undo the cultural changes that have placed all this pressure on our genomes, but it is realistic to hope that we can come to a better understanding of how our genes interact with one another and the environment. This is not a book about prescriptions. For all the anguish in our lives, arguably humans have never experienced as much happiness as they do today. For all the discomfort of old and middle age, we live longer and have immeasurably more comfortable lives than ever before. This book is fundamentally an attempt to address an imbalance in the way that nature and nurture are presented as one-or-the-other causes of disease. Every publication of a gene for cancer or a gene for diabetes or a gene for aggression raises expectations for a fix, be it genetic manipulation, or a simple over-the-counter drug; some might even desire a eugenic breeding program. No, genes are not instructions engraved in the tablet of our DNA determining this or that. They are just sequences of letters that orchestrate tendencies, and we ought to embrace their variety.

This is the idea we return to in Chapter 8, "Genetic Normality." Why is it that some people are tall and some short; some are dark-skinned and others fair; some people (and even some Vice-Presidents) seem to have a permanent scowl while others have bright, welcoming smiles? There is ample variation for the dimensions of psychology, such that some of us are introverted yet open to new ideas, some are conscientious and emotionally stable, and others are disagreeable. Little is known about the genetics of such attributes, other than that they do run in families, so it is really only a matter of time before we begin to find the genes. Already, though, we can be sure that they will tell a story much like those related to disease, a story of complexity and of interactions.

Normality is a distribution, not a category. Distributions have extremes, and those extremes are as much a part of life as the fact that most individuals occupy the middle. Every single one of us is extreme for some human attribute, because we inevitably have some unique constitution of genetic variation. To be human is to participate in the extraordinary diversity of flavors in the human gene pool, a diversity that may be responsible for illness and frailty but deserves equal credit for human creativity, beauty, and accomplishment.

The ideas in this book have incubated over a period of 15 years, and I would like to collectively thank and acknowledge all of the many colleagues who have encouraged, supported, and criticized along the way. Of course that particularly means everyone in my research group, whom I owe more gratitude than I can express. A few people have been particularly important over the past year or two: Karl Leif Bates and Beth Weir read drafts of most chapters early on and provided me with the confidence to continue, while David Goldstein was also a rock of encouragement at a critical juncture. Their task was made easier by the fact that a wonderful puppy named Razzie got me out of bed for walks and morning writing sessions. Amanda Moran and Tim Moore at Pearson have taken a punt on the book, and I cannot thank them enough for that and for allowing me to be part of the new science imprint at FT Press. Russ Hall has been the perfect editor for me; any deficit in his powers of persuasion is simply a product of the fact that my own degree of stubbornness is highly canalized. The book covers a huge space of biomedical literature, and since I mainly study flies rather than specific diseases, it is inevitable that errors of fact and interpretation have crept in: Responsibility for these is completely mine. I hope that the overall message nevertheless seeps through. Finally, because she is my antidote to all the complexity in the world, I dedicate this book to my wife, Diana.

Greg Gibson
Raleigh, December 2007

1

The adolescent genome

genetic imperfection Disease is a normal and inevitable part of life that arises from the way that organisms are put together.

unselfish genes The way that different flavors (*alleles*) of thousands of genes work together establishes how an organism looks or behaves, or how healthy it is.

how genes work and why they come in different flavors
We all differ from one another at millions of places in the genome.

three reasons why genes might make us sick Rare alleles that have a large impact, common alleles that have a moderate one, or hundreds of alleles with very small effects can all contribute.

a unified theory of complex disease The combination of rapid human evolution and recent cultural change has pushed us out of a genetic comfort zone, predisposing many more people to disease.

the human genome project Public and private efforts have jointly produced a complete sequence of the human genome that lays the foundation for a century of medical research to come.

genomewide association The scalpel that will be used to isolate most of the major disease susceptibility alleles for complex disease of the next few decades.

genetic imperfection

Of all the paradoxes in the world, surely one of the most absurd is that the very same genome that gives us life inevitably also takes it away. Even when they aren't killing us, our genes are generally making existence more difficult than seems absolutely necessary. Very few people escape this world having avoided a bout with cancer or diabetes or asthma or depression, and those who do often end up too senile to remember much of the journey anyway. What good reason could there possibly be for so much suffering and disease?

Maybe there is no good reason, other than that genetic disease is an unavoidable byproduct of the way organisms are assembled; **disease arises because humans, like all other species on the planet, are an unfinished symphony. Perhaps we are even more unfinished than most, thoroughly out of equilibrium with the modern world, and even a little bit uncomfortable in our own skin.** In short, we possess an adolescent genome.

This notion may seem counterintuitive, because we are so conditioned to think in terms of perfection. A simplistic way to think about biology is to imagine that every species is perfectly suited to whatever ecological niche it occupies. Its genome has evolved to ensure that each individual is made to be as close as possible to the optimum shape and set of functions that a perfect member of the species would have. Adaptation to a dragonfly is having exquisitely refined lace wings, to an orchid it is pitching the lips of its pouting petals at just the most attractive angle, and to a human it is whatever it takes to live a long and comfortable life. Maybe no one individual is ever truly cast as the ideal that defines the species, but all approximate the optimum.

If an individual doesn't quite define perfection, it is either because optimality actually comes in a variety of shapes and sizes, or because forces are conspiring against the person. Debating whether humanity is more closely realized in the form of Colin Powell or Tiger Woods, Jennifer Lopez or Hillary Clinton, we would no doubt agree to disagree on what attributes are desirable in a person. We would, however, likely find common ground when it comes to health, concluding that some not-so-optimal types of genes floating around make us hypersensitive to pollen, push us to eat too much, or make us prone to mental illness. So the question is, why are such bad influences tolerated in the gene pool?

As the book unfolds, we will look closely at six different types of disease, each given its own chapter. It is first necessary to lay the foundation, so I have three goals in this opening chapter: first, to disavow you of any sense that there is such a thing as a "disease gene;" second, to lay out the general theory of complex disease that I enunciate as the book unfolds; and third, to explain how contemporary geneticists go about finding the genes that influence susceptibility to illness.

unselfish genes

Telling someone that she has the gene for Parkinson's or the gene for restless leg syndrome is a bit like telling her that her house has termites or sits on a toxic dump. It implies that her misfortune is that she has something that most people don't have, and further that all would be well if only she could get rid of the termites or toxins.

Genes are not like that, though. They are not things that some people have, and others do not. Approximately 23,000 genes are in the human genome, and all of us have pretty much the same number, give or take a few dozen. What we actually have are different flavors of genes. The technical term for a gene flavor is *allele*, pronounced *ah-lee-el*: Whenever you read the word "allele," think of chocolate and vanilla ice cream. Alleles are different versions of the same gene, just with different spelling and slightly different function.

In fact, in many cases, when a gene is associated with a disease it is because the gene is in some way broken or missing. Just getting rid of the gene would not help. A better house analogy than termites and toxins might be damp foundations, or cheap window frames. The house is basically the same as everyone else's, but problems arise because it just wasn't built as well as it should have been. Generally in such cases, many other things also are likely to go wrong and in this sense, too, the analogy with complex disease is improved.

Similarly, it seems that almost daily we read proclamations that scientists have discovered the gene for stroke or the gene for homosexuality. In almost every case, what they really mean is that the scientists have discovered a particular variant of a gene that slightly increases the likelihood that some people will suffer strokes or prefer their own sex. Sometimes the headlines replace "the" with "a," which is definitely better but still conveys the impression that the purpose of such genes is to cause the

disease or trait. Actually, the genes universally promote what we collo-
quially refer to as normality. They come in different alleles, and under
some conditions particular alleles promote disease, or conditions we
choose to label abnormal.

Contemporary genetic research is focused on finding these alleles
and is as much about basic understanding of what they do as it is about
finding cures for specific diseases. This is because there is little prospect
of finding new cures for cancer until we understand why tumor cells
grow out of control in the first place, and the next drugs for treatment of
depression won't arrive until we appreciate what is wrong deep inside
the brains of the chronically sad. This makes sense if you consider that
most of us would prefer that our automobile mechanic understand how
the engine works, rather than just try the same old fixes he's always used
in the past.

The advantage that a mechanic has is that humans made cars, so we
know not just what every part does but also what its purpose is and how
it interacts with all the other parts. Biomedical researchers now have a
pretty complete parts list and a fair idea of where each part goes, but
there is still much to be learned before we know what all the parts do and
how they fit together to make a healthy person.

Much genetic research involves pulling apart and putting back
together model organisms that we can manipulate, like mice and rats and
zebrafish, and even flies and nematodes. Increasingly the tools are at
hand to do it with humans directly—at least, the pulling apart bit. Also,
for just about every gene, somewhere in the world there is a person with
an allele that does not work, and many thousands of these are responsible for rare syndromes. They are teaching us a lot about how things function, but for the most part don't explain the common diseases that afflict
us all.

To this end, a parallel mode of genetic analysis is much less familiar
to most people and yet influences all of our lives on a daily basis to a
much greater extent than the genetics that we learn in school. Variously
referred to as *quantitative genetics*, or by phrases such as *complex
disease*, *multifactorial trait*, or *polygenic disorder*, it is the study of how
common variants in many genes interact with one another and with the
environment to produce the biological variation that surrounds us. Genes
are fundamentally interactive entities, working together, adjusting to the

environment around them, molding organisms but not determining their destiny. For anything the least bit complicated, it truly takes a genome.

Most of the differences between species are of this type, as are the attributes that make us unique, from body shape and facial features to metabolism and even aspects of temperament. So too are the diseases that touch every one of us directly or indirectly as they afflict friends and family: cancer, diabetes, cardiovascular disease, asthma, and depression. The language associated with quantitative genetics switches from the imagery of control, determination, and causation, terms popularly associated with genetics, to the less strident tones of susceptibility, influence, and contribution. This book is predominately about the genetics of complexity.

Perhaps another analogy might make the distinction clearer. All of us are probably painfully aware of the impact that one individual can have on a business. If the CEO, or CFO, or CSO, or Director of IT, or head housekeeper for that matter, stops working or starts making bad decisions, the company can deteriorate rapidly. Yet it is the more subtle failings or distractions of multiple employees that most often disturb the health of the company even in good times. Two co-workers are going through a divorce, a supervisor is having an affair, the junior VP for marketing is caring for her ailing mother, and one of the bookkeepers has repetitive strain injury. Nothing is particularly unusual about any of these circumstances, and each of them is almost to be expected in even a moderate-sized group of people. For the most part organizations can and do deal with them, but mix them together in certain combinations and pretty soon potential dissipates, opportunities are lost, maybe employees start leaving, and things can fall apart. Such is also the fate of our genomes: Genes are ultimately individuals that have to work together, but they're not perfect, and sometimes the pieces just don't mesh.

Far from being selfish robots, genes are in fact little molecular existentialists. Contemporary molecular biology is about relationships and networks. It is the context within which a gene is used that defines what it does and what it is. Sure, certain genes are essential for the development of the eye or the heart, but these same genes do other things in different contexts. Think not of genes as dictators, but rather as a parliament of constituents—a parliament that on the whole does a pretty decent job, but sometimes messes up, with dire consequences for the health of the organism.

how genes work and why they come in different flavors

Even if you haven't asked yourself why it is that genes makes us sick, perhaps you have wondered why it is that your sister has legs up to her ears and piercing blue eyes that haven't been seen in the family since Great-Aunt Bessie, while you seem to have inherited a horrible mix of dad's stockiness and mom's frumpiness? And what's up with your brother's moroseness: Where did that come from?

It is not much of an explanation, but the straight answer is that **genetics is a lot more complex than the idea that there's a gene for every trait.** Most traits, or attributes, are regulated by many genes, not just one. Furthermore, while it is a nice abstraction to suppose that genes come in normal, or good, versions and mutant, or bad, ones, the reality is that there are always multiple different flavors of normal. The gradation from the most common allele to various types of normal alleles to abnormality is continuous. Just having certain alleles is insufficient to predict whether a person will get a disease.

Crucially, too, the environment has a pervasive effect on the way our genes function. "Environment" means much more than the temperature outside or the nutritional content of the food we eat. It also includes influences as diverse as a mother's health during pregnancy and the pressure that peers and society put on us to behave in certain ways. As we shall see, in many cases environmental interventions are likely to have a much greater impact on public health than pharmaceutical ones. Unfortunately, most of us find it easier to pop a pill than to buck a social trend, so drugs are likely to have an ever-increasing role in disease control.

Without going into any mechanistic details, it is helpful to recognize that genes function on two levels, the biochemical and the biological. The biochemical is hidden to most observers, and therefore typically excluded from general conversation. The biological is what we actually see.

Each of the 23,000 or so genes scattered along our chromosomes encodes the information to perform a specific biochemical function. Less than a third of these genes function in every cell in your body to provide the basic building blocks and to generate energy—they are the bricks and mortar, if you like. Another third of our genes makes every one of the hundreds of different cell types in your body different. Neurons need proteins that process electrical signals, muscle cells are full of actin and myosin that make them stretch and contract, and white blood cells carry around the components of your immune system. These are the doors and windows and furniture and appliances. The final third of our genes is responsible for regulating which genes are used when and where and in what amount. Turning on hair keratin in your pancreas wouldn't be good, and light receptors have no place in your heart, so development and physiology are highly regulated processes. These genes are the architects, foremen, and designers.

We hear and read about genes for cancer and for autism, or are given to believe that there is an aggression gene or a blonde hair gene. The reality is that these attributes are many steps removed from the molecular functions that the genes perform. If a gene contributes to cancer, it is because it normally performs a role in making sure that the right number of cells are produced at the right time and place. The reason there may be a genetic contribution to spirituality is not because some genes function to ensure that we have a belief in God, but rather because there are genes that affect how the neurons are wired together and the strength of signaling across synapses.

Fly geneticists like to name genes after the way flies look when the gene is mutated. *Antennapedia* flies have legs on their heads, *technical knock out* ones fall over when you bump their heads, and *shaven baby* embryos don't have any hairs. It is an amusing, but unfortunate habit, because it reinforces the notion that there are genes for traits. Time after time it turns out that the same gene does completely different things in different contexts. A favorite example of mine is *staufen*, which is required both for sperm development and for memory. It is not that male flies think with their penises, but rather that both of these attributes turn out to depend on a biochemical process called intracellular RNA localization, which *staufen* is involved in. Almost without exception the biological functions of genes are not written in the DNA, but rather emerge from the network of biochemical interactions within cells, and in turn the manner in which cells work together to build tissues and organs.

It follows that the reason we are all a little different from one another is because these interactions occur between ever so slightly different copies of the genes. Each gene comes in multiple different flavors—I mean, alleles—that have cropped up during the evolution of the species. These different alleles have their origin in the process of mutation, which is basically what happens to genes when you leave them out in the sun or exposed to poisons.

Mutations are ultimately the source of all things good, but for the most part are harmful, tending to break genes. Every one of us has a few mutations that neither of our parents had, simply because mistakes are made every time the genome is copied. (But don't get too upset about this: The error rate is only about one in a billion letters in the DNA. Most of us would be thrilled to make a mistake only once in every hundred times we do something.) Mutations are also so plentiful that we all carry several of them that would kill us if we got the same one from both parents.

Mutations are so plentiful in fact that there is no way that natural selection can possibly purge them all. Obviously alleles that would tend to kill a person will not generally last long in the gene pool, and similarly ones that would tend to make us sick should not fare well either. But all new mutations are extremely rare when they appear, and nature has bigger fish to fry. It is more concerned with common alleles that affect the fitness of a large percentage of the population, so the fate of new mutations is largely governed by chance. Consequently, some mutations manage to drift around for a while and can even become reasonably common before they start having a noticeable effect on public health. The process is called *mutation-selection-drift balance*, which is a fancy way of saying that a lot of bad things happen to genomes, and evolution deals with them, but it is so busy that some of the bad things hang around for a while.

Some mutations are also good for you. Maybe they offer protection from diabetes; maybe they make a person more fertile. These tend to be favored by natural selection, but before they become the standard allele, they necessarily share real estate in the genome with the original allele. Typically it takes thousands of generations for one allele to replace another, so in the meantime you have variation. Sometimes the new allele will be better under some conditions, while the ancestral one is better under others. Maybe they have different effects in men and women, or in rural and urban settings. In such cases, geneticists speak of *balanced polymorphisms*, the classic case being sickle cell anemia, which is bad under some circumstances but protects a person from malaria in others.

You will also see it argued that many of the bad effects are actually offset by some absolute good that they do. Perhaps at a different stage of life they are sufficiently beneficial that natural selection overlooks their contribution to disease. Or perhaps at some earlier phase of human evolution they were the right gene in the right place at the right time. It is easy to get carried away with devising clever stories along these lines. Some, particularly in the domain of psychology, are even tempted to postulate that promoting disease is in itself advantageous to the selfish genes, but it really stretches credulity to suppose that there is some benefit to having genes that make us suicidal. We won't go down that road. Rarely is it necessary anyway.

It turns out that as species go, humans are actually among the least variable, at least at the level of their DNA. Nevertheless, the average person has a few million differences between the copy of his genome received from his mother and the copy received from his father. Somewhere among all those differences are the genetic variants that are responsible for all genetic diseases, but no more than a couple dozen have a big enough impact on any particular disease for us to have any hope of finding them. Finding a few dozen out of a few million is a genuine needle-in-a-haystack problem.

three reasons why genes might make us sick

The upshot of all this is that there are basically three ways that genetic variation can influence disease susceptibility. These are called the *rare alleles*, *common variant*, and *small effect* models. I will briefly describe each and then in the next section present a unified framework that iterates throughout the remaining chapters.

The simplest model is that a disease can be traced to one badly disrupted gene. This is pretty much the case for cystic fibrosis, and for thousands of other rare conditions. Around 1 in 100 of us carries a mutated version of the *CFTR* gene without any ill effects, but if two carriers marry, their children have a 1 in 4 chance of getting both bad copies and consequently having cystic fibrosis. The incidence in the general population is only about 1 in 10,000, most of which is due to a few mutations that have been around for centuries, but actually hundreds of other mutations can be found in the gene as well. Whether the disease is so severe that it claims the life of an infant, or mild enough that a person can live to adulthood and maybe receive a life-saving lung transplant, is in part a function of which mutations they have, in part of the rest of their genome, and in part their upbringing.

Single genes can also cause diseases in other ways. Muscular dystrophy is often due to a gene, *dystrophin*, that is so big that it picks up mutations often enough that most new cases arise in the individual who has the disease. Another small set of genes has an odd feature that makes them mutate at an unusually high rate, leading to the paralysis or ataxia observed in Fragile X syndrome and Huntington's disease. For the most part, though, single gene diseases are rare.

Large-effect mutations also do not generally explain common diseases, those seen in five percent to ten percent or more of people. Really the only way they could is if there were hundreds of genes that cause a syndrome that we choose to think of as a single disease. Schizophrenia might be in this class, as might the wide spectrum of cardiovascular conditions that lead to heart attacks and stroke. It is possible that these rare mutations interact with one another, so that a person needs two or three of them in any condition to be predisposed to the disease. Unfortunately geneticists have not yet devised a systematic way to discover such mutations.

Currently the most popular model is called the common disease-common variant, or CD-CV, hypothesis. It is the idea that if there are diseases found in ten percent of the population, then there ought to be alleles at about the same frequency that are found in these people, but not in "normal" people. This sounds reasonable enough, so millions and millions of dollars are being spent in pursuit of these alleles, each of which contributes about five percent to ten percent of the risk of illness. So far, Crohn's disease, an inflammatory bowel syndrome, is the poster child success story, except that it is not actually a common disease. However, ten or so genes have been discovered that contribute to Crohn's, each with correspondingly common risk alleles. Diabetes and prostate cancer also show signs of following the CD-CV model, but the jury is out on whether this will really be a common explanation for disease.

The third possibility is that hundreds if not thousands of different genes—each with rare or common alleles that have small, barely detectable effects—contribute to each common disease. To some extent this is the default model when all other models fail, but it is beginning to look like it is going to be the predominant explanation. The trouble is that this model doesn't really explain why diseases are discrete. Height, degree of extraversion, memory performance, and probably most human attributes are thought to be influenced by hundreds of genes, but they show a continuous gradation from short to tall, shy to outrageous, and forgetful to prolific. **So why should there be people with disease and people without disease, if hundreds of genes are involved?**

A somewhat technical explanation for this is that there is a threshold of liability—in other words, a tipping point from health to sickness whenever you have a little more of something than is normally tolerated. Most people are pretty similar genetically, having average levels of whatever it is. They have some genes that increase the attribute and some that decrease it, but generally not an excess of either. However, inevitably a few outliers will have considerably more of the increasing or decreasing alleles, enough to send them beyond the threshold into the valley of illness.

a unified theory of complex disease

An added quirk is that there likely are mechanisms that ensure that as few individuals as possible exceed the threshold, even when they have more than their fair share of the risky alleles. This phenomenon is known as *canalization*. It says that not only do species evolve so that most individuals resemble one another, but they have also evolved buffering that ensures that everyone is "normal" despite the slings and arrows of outrageous fortune that life throws at them.

Next time you trap a mouse, count the number of whiskers: Almost certainly there will be 17 or 18 on each side of the snout. Actually, my dogs also have this number of whiskers, but that may just be coincidence. This number of whiskers is very stable, unless the mouse happens to have a *Tabby* mutation, in which case on average it will only have a dozen or so whiskers. The catch is that the "or so" can be as few as 7 and as many as 20. Observations such as this are often seen when developmental circumstances are perturbed. Not only does the average appearance change, but it also becomes much more variable.

It seems than that normal buffering mechanisms fall apart when the genetic system is pushed too far away from the optimum. Translated into the realm of disease, **the idea is that the modern environment that humans have constructed has taken us out of the buffering zone, and left us more susceptible to perturbations that result in disease.** It is, however, much easier to describe what canalization is than the mechanisms that produce it. This is partly because we don't really understand the mechanisms, and partly because they are usually addressed in mathematical and statistical equations.

The essence of these equations is that stability arises through the deeply interconnected web of interactions among genes. If I were to give you 100 pieces of string and ask you to make a carrying bag, the simplest thing you could do would be to tie them all together at both ends, resulting in a sling. This would be fine for carrying around tennis balls, but somewhat disappointing if you tried to use it to carry your loose change. A slight improvement would be to divide the strings into two groups, and lay two slings perpendicular to one another. If you had time, you could weave the strings into a cross-hatching cloth, and by adding reinforcing strings at different angles you could make this web even stronger. Such a cloth would be able to hold heavy objects that distort it and to absorb breaks in a few of the strings.

Genetic networks are similarly structured as interacting linkages that together form a tighter, more coherent whole than would be produced simply by adding together bits and pieces. But the whole inevitably has holes, particularly when stressed, and these holes lead to disease.

Now think about some recipe you used to love to make as a child. Let's say your favorite ham and cheese omelet, or if you were unusually adept in the kitchen, a soufflé. When you were a child, you probably stuck pretty close to the recipe, knowing that so long as you balanced the amount of ham and cheese you added, the omelet would turn out nicely. Then you went away to college and went through a phase of not eating breakfast or stopping for a McBiscuit on the way to work, and now you've forgotten the exact recipe. You think you have it right, but every other time you make one, the kids get a pained look on their faces and spit it out. There's probably something wrong with the number of eggs you are using or the amount of milk. Or maybe it is because you are using an electric stove instead of gas, or the eggs where you live now are a different size than those where you grew up. It's frustrating, but you just can't recapture the magic of the old combination.

In this metaphor for the origins of complex disease, the recipe stands for the genetic program for healthy development, growing up and changing the recipe stands for genetic evolution, and switching cooktops stands for environmental change. The key is that tens of millions of years of genetic evolution devised canalized systems for regulating the amount of glucose in our blood; the balance of immune response to bacteria, viruses, and parasites; and the way the chemicals signal in the brain. These systems were well able to absorb normal fluctuations, without exposing too many individuals to disease. But humans are an incredibly young and rapidly evolved species, and we have completely changed our environment in the past century. This pushes us—as well as many of our domesticated companion animals that get similar diseases—out of the buffered zone, exposing genetic variation that may never have had an effect in the past.

So while it is convenient to assume that humans are close to some optimal design, we have not actually been around for long enough to allow the genome to make fine adjustments that ensure that most people are buffered from disease. Humans are without a doubt a long way from any such equilibrium. We shared a common ancestor with chimpanzees just five million years ago, and with *Homo erectus* cavemen just a million years ago. As a species, *Homo sapiens* has been in existence for just 140,000 years, somewhere around just 10,000 generations. The flies sitting on the fruit salad at your barbecue have likely been around as a species for 100 times as many generations.

Perhaps it wouldn't matter so much, except that we're also a really, really different species in so many ways. We're just beginning to explore our novel world. From the Arctic to the Antilles, and from Newfoundland to New York, humans are re-creating their niche, putting pressure on the gene pool to deal with all kinds of extremes. We live longer than our close ancestors, consume strange diets, walk upright with a funny pelvis, have babies with big heads, share our homes with a menagerie of animals, and cope with really complex social settings. If you feel stressed at times, imagine things from the perspective of the genes that helped us get here.

The point is that recent human evolution has required substantial changes in our genetic makeup, disrupting genetic relationships that had evolved over millions of years. These changes have left us exposed. Like an adolescent still growing up and trying to come to terms with a constantly changing world, we're just a little uncomfortable with who we are. Presumably we'll get to a more comfortable genetic place, but not for a few more hundred thousand generations.

the human genome project

Let's turn now to the issue of how geneticists study the origins of disease, beginning with something called the Human Genome Project. This is an effort to identify and describe the function of every one of the genes in the human genome, particularly those related to disease. Early on, there were some naïve expectations that just by sequencing a genome, the genes would be obvious and within a few years we would have cures for all the major maladies that afflict citizens of the developed world.

It hasn't turned out that way, for good reasons, but the technical accomplishments have exceeded expectations, and it is doubtful that anyone foresaw the direction that genome science would take. The first announcement of a draft human genome sequence was greeted by President Bill Clinton as a step toward a closer understanding of God's design. Less spiritual observers saw it as a step toward diagnostics and interventions for hundreds of diseases. Cynics saw it as yet another example of scientists' hubris in throwing hundreds of millions of dollars at a problem without solving anything. My sense is that, like man's walking on the moon, it is an achievement that serves as an identifiable landmark in the emergence of a new domain of human endeavor, but will eventually be seen as just another small step along the human journey of self-perception.

There were actually two genomes sequenced—one by an international consortium that was financed by public money, and the other by a commercial enterprise known as Celera Genomics. It is legitimate to ask why hundreds of millions of taxpayers' dollars were spent on a project that turned out to be doable by private initiative. There are many answers to this question. One is that Celera might never have started without the incentive provided by the public effort (and similarly, the public effort would not have finished so quickly without being pushed by Celera). Another is that there were legitimate reasons to believe that the strategy adopted by Celera would not work, whereas the public approach was guaranteed both to work and to provide useful information as it progressed.

The two projects took what we might refer to as MapQuest and Google Earth strategies toward sequencing the human genome. Suppose that you are asked to come up with a brand new atlas of the United States, complete with street names and house numbers. Most of us would probably start by employing someone in each state and charging them with the task of mapping out the major cities and highways. Lake Tahoe and Fresno would be placed on the atlas as the cartographers radiated out from Los Angeles and San Francisco, eventually linking up with Reno and Las Vegas. The approach would be painstaking and slow, but for most intents and purposes, guaranteed to be accurate, and the drafts could be used even before the final version was available. This is what the public effort did: Each chromosome was assigned to a major sequencing center, and the consortium put the pieces together over a period of five years.

By contrast, the maverick visionary behind Celera, J. Craig Venter, decided to do the equivalent of renting a satellite to take hundreds of millions of photographs. Every piece of land would be present on at least ten of the photographs, and a massive supercomputer was programmed to find the bits of similarity at the edges, assembling the complete atlas simultaneously based on overlaps between adjacent photographs. The process was fast and relatively cheap, but you might imagine that all those Main Streets and repetitive cornfields in the Midwest would confuse the alignment of photographs, making for some odd distance estimates, and that bits of New York might turn up in the middle of Philadelphia by accident. However, the Celera people were clever enough to devise ways around these problems, and their atlas of the human genome turned out to be just fine. It also turned out to be Craig Venter's own genome!

Bear in mind that an atlas is just a set of guides: It doesn't tell you where steel is manufactured, where cotton is grown, or who lives at 286 Magnolia Lane. For that we need classical genetics, bioinformatics, and molecular biology. The biggest emphasis right now, though, is on comprehending the variation at each of the positions in the atlas. This is the quest to find the tens of millions of places in the genome where we all differ, and to work out which few thousand of these differences are associated with disease and behavioral, physiological, and physical variation.

We don't need to get into the gory details of how the genetic code can possibly hold the secret to life; suffice it to say that it consists of four letters, A, T, G, and C, strung together in long molecules of DNA. Each human gene is made of something like 10,000 of these letters, and there are around 1,000 genes per chromosome. The sequence of these letters specifies the nature and function of each gene; sequences can vary among individuals in three basic ways. *Single nucleotide polymorphisms*

(SNPs) are positions in the genome where two or more different letters might be found if you compare two people. *Indels* are insertions and deletions, usually of just one or a few letters. Thus, comparing the sequence AATGCGCA with AGTGCGCCA, it appears that there is an A/G SNP at the second position, and an insertion of an extra C two bases from the end of the second sequence. The third class is *copy number variation (CNV)*, which is much larger insertions and deletions of thousands of bases at a time.

Ultimately, change in the sequence of letters translates into susceptibility to disease or blue eyes or a cheery disposition. Remarkable as it may seem, a person who has an A instead of a G at position 102,221,163 on chromosome 11, may be born with a mild heart defect. This insight leads to the idea that if we could sequence a person's genome, we could maybe work out what diseases that person may be likely to get. Venter has written a book that does just this for himself, but it must be stated that our ability to interpret sequence differences is primitive, and like a Shakespearian play or a T. S. Eliot poem, the genetic words will always be subject to interpretation.

So what has the sequencing of the human genome achieved, and when should we expect to see some impact on medical care? The achievement is that we now have the solid foundation upon which twenty-first century biomedicine will be erected. To see this, think about the analogy of the genome sequence as a road atlas once more. Prior to its completion, molecular biologists were simply working with the obvious features in the genomic landscape, or laboring painstakingly to find where they were headed. Now they know exactly where the residential areas are, where to find businesses or manufacturing sectors, and what type of agriculture is carried out where. They have the street names and addresses for most families and can readily find the government regulators.

To identify what is going wrong when a genetic disease occurs, though, they need to be able to peer inside the houses and offices. Sometimes the cause of a problem is obvious, as if the roof were missing. More generally, though, subtle problems get in the way. The most interesting cases don't involve a single mutation, but rather the accumulated effects of many regular, everyday variants in the genome. These variants are behind diabetes, asthma, and depression, and they take advanced statistical and computational procedures to find.

genomewide association

Until the middle of 2006, the search for new genes that influence disease was pretty much restricted to studies of extended families. Typically, geneticists would identify pedigrees in which a particular type of cancer or heart disease was unusually common and look for parts of the genome that affected individuals have in common. This approach, called *linkage mapping*, has been the main method for finding single gene disorders, but has had limited success for more complex diseases.

Your parents have between them four copies of every gene. You have two of these, and your children each have a 50-50 chance of receiving each one. Suppose now that you, your father, and two of your three kids have a heart murmur, and both of these kids received the same allele from you, which is also the one you got from your Dad, while their sibling received the other allele. So, four out of seven members of the family have the murmur, each of whom has the same allele of the gene. Something fishy is going on, and you would likely conclude that the level of correspondence between having the allele and having the disease is unlikely to be due just to coincidence. You would suspect that the allele actually causes the murmur.

However, since there are thousands of genes and millions of families with murmurs, that level of coincidence is bound to occur occasionally. But if geneticists find a similar correspondence in dozens of even bigger pedigrees, their confidence that the particular allele of the gene actually causes or at least contributes to the murmur increases. With enough data, the correlation between the gene and the disease does not have to be 100 percent. As a result, it is also possible to detect linkage between regions of the genome and complex diseases where each gene only has a small influence on the disease. On this basis we know, for example, that a dozen or so places in the genome influence type 2 diabetes. For reasons we needn't concern ourselves with here, those places typically stretch over perhaps a tenth of a chromosome, or hundreds of genes. So they do not pinpoint the problem.

To get around this, the field has now turned to a revolutionary approach called *genomewide association mapping*, or *GWA*. Instead of looking in families, geneticists now look at unrelated individuals drawn from an entire population. Two companies, Affymetrix and Illumina, have manufactured little gene chips with up to a million common genetic differences printed on them. These markers stand as proxies for the tens of millions of places in the genome that are different among people. For less than $1,000 a pop, geneticists can now effectively measure what a person's genetic constitution is, almost as if they were determining the sequence of the person's entire genome.

For a few million dollars geneticists can go out and compare the genomes of 10,000 people who have a disease, with the genomes of 10,000 people who do not have it. If the frequency of the A at position 102,221,163 on chromosome 11 is 29 percent in people with a heart defect, but only 19 percent in people without the defect, then after appropriate crunching of the numbers we can infer that this site is contributing to the problem. This is a gross oversimplification; all sorts of possible alternative explanations can be made for such a difference. But if another group replicates the result in an independent sample (often from another country), then confidence that the gene is involved in the disease shoots up still higher.

It turns out that this approach is a sufficiently fine genetic scalpel that it actually leads us to one or a few genes involved in the disease. Genomewide association scans for disease will be to human genetics what the microscope was to nineteenth century biology, and what they are telling us is rightly the subject of the remainder of this book.

2

Breast cancer's broken genes

cancer of the breast   Breast cancer is the most frightening source of mortality for women but is rarely genetic, and early detection of the disease offers the best protection.

broken genes, broken lives   The genetic component of cancer is because genes break during your own lifetime, more than because you inherit already broken genes.

epidemiology and relative risk   The incidence of breast cancer varies across the globe in ways that implicate modern lifestyle as a major culprit.

brakes, accelerators, and mechanics   Three types of things go wrong in cancer that affect the control of cell growth and division.

familial breast cancer   *BRCA1* and *BRCA2* are the best known genes involved in breast cancer, having a major effect in some families, but they account for a only small fraction of susceptibility in the general population.

growth factors and the risk to populations   New genetic methods are finding common genetic variants that increase your risk of breast cancer ever so slightly.

pharmacogenetics and breast cancer   How you respond to drugs such as tamoxifen is also influenced by your genetic makeup.

why do genes give us cancer?   Until recently, cancer was not a major cause of human ill health and mortality, so there has been little pressure to rid the genome of hundreds of variants that predispose to cancer.

cancer of the breast

The list of famous women who have had breast cancer contains some extraordinary pairings that on the face of it suggest something other than coincidence: both of Charlie's brunette Angels, Kate Jackson and Jaclyn Smith; the American singer-songwriters Carly Simon and Sheryl Crow; similarly the Australian pop divas a generation apart, Olivia Newton-John and Kylie Minogue; the classical actresses Bette Davis and Greta Garbo; and the diametrically opposite feminist personalities Suzanne Somers and Gloria Steinem. But at a lifetime incidence of one in nine contemporary women having breast cancer, such a list is not unusual, shocking as it is.

Less well known is the fact that lung cancer is even more common in women and, given the incidence of cigarette smoking in teenage girls and twenty-something women, is likely to increase. Heart attacks are a still greater source of mortality, but breast cancer is undoubtedly the lightening rod for women's health issues.

Chances are, everyone knows someone who has had breast cancer. If they are friend or family, we will have shared some of the ordeal, experienced secondhand some of the pain and anxiety. In all likelihood and without realizing it we all also know women who have survived. A generation ago, breast cancer carried such a stigma and was so closely associated with a death sentence that it was kept as private as possible, almost a taboo topic of discussion. Some people may have felt the disease was contagious; others may have attached shame to the symptoms, as if contracting cancer is a woman's fault—as if conquering the physical illness were not enough to deal with.

The reality today is that the vast majority of patients will survive. Catch a small tumor early enough, and the improvements in surgical procedures, drugs, and more focused radiation therapies almost guarantee recovery. Five- and ten-year survivorship rates are on the up, and we have started to hear that refractory cases (those stubborn or resistant to control) may not be curable but are certainly manageable.

Read any of the self-help books on the topic, and you actually get the impression that surviving breast cancer can be an extraordinarily life-affirming experience. One of the best is Barbara Delinsky's *Uplift*, a compendium of short testimonials on how to use a positive outlook to cope and overcome. Many are the families that have been brought closer together by sharing the ordeal. Many are the women who find an inner strength they never knew they had, or who reevaluate what really matters in their lives. Many are the new friendships that are made through membership of the sisterhood.

The book is full of vignettes ranging from the prosaic to the profound. There's advice for padding bras and purchasing wigs, for organizing drainage tubes after a mastectomy, and for being creative with the little tattoos that oncologists use to guide their X-ray machines. There are funny stories of flat-chested daughters welcoming formerly buxom mothers to their world, and of sadly drooping older women finding new confidence in the simply fabulous shape of an implant. There is the man with a lump in his breast who presents to a clinic where he is asked to fill out pages of forms documenting the recent regularity of his menstrual cycling or early signs of menopause. And most shocking to me was the sense that so many women are afraid to tell their husband for fear that he will think less of them if they have a breast removed, as if men are that superficial!

Early detection of breast cancer generally begins with a mammogram. Next time you are in a crowd, perhaps at a baseball game or a concert, contemplate the fate of the middle-aged and older women around you. One in nine will be diagnosed with breast cancer over the next 30 years. If 1,000 of these women were to attend a clinic that day, it is almost certain that several new cases of the disease would be detected. For the most part, the diagnosis would be early enough for physicians to intervene, and the prognosis would be excellent.

Perhaps in a few cases the woman may have been able to detect a lump by self-examination; or may have experienced unusual tenderness or discharges from a nipple, that she put down—understandably—to stress or growing older. Some will have delayed making a doctor's appointment, out of work commitments or denial. One thing we do know, though, is that the more advanced the tumor, the larger it is, and the more time it has had to spread out of the breast, the worse are the prospects.

Of the 1,000 women, statistics indicate that about 70 would be called back for follow-up consultation. A mammogram is a low-intensity X-ray of the front and side of the breast, designed to reveal abnormal densities of cells. For the most part, these turn out on closer examination to be just a normal aspect of the breast tissue. In fact, particularly in women under the age of 40, there is typically so much density that it can obscure tumors that are present: It is thought that about ten percent of early breast tumors are missed on mammograms. More accurate procedures such as ultrasound and magnetic resonance imaging will quickly tell the doctors that 60 of the possible cases actually had a benign explanation.

Of the ten out of each thousand women whose situation looks dire enough that they are next referred for biopsy, only three or four will in fact have a cancer. Two-thirds of these will be early enough that they will almost certainly be cured. So the screening of 1,000 women will likely save the life of a couple of them, at the expense of several months of distress for as many as 100 others, while it may have missed a case as well. There is also the issue of whether the energy of the X-rays themselves can induce cancer. While this risk is miniscule for a single scan, it is incremental, so annual mammograms starting in the thirties are no longer recommended. One every few years starting at menopause seems a reasonable compromise.

broken genes, broken lives

Cancer touches every one of our lives in one way or another, seeming to show little care or concern for whom it attacks. There does not seem to be much we can do about preventing it, aside from eating more carefully and avoiding smoking. Perhaps because of the visceral nature of the disease that so clearly arises inside our bodies, unlike an infection transmitted by sex or perhaps an insect bite, cancer is commonly regarded as a genetic problem.

We're told it is often hereditary, and this is a source of particular worry when common experience reveals that a couple of relatives in the extended family have had cancer. But since it is so common and so diverse, in most cases coincidence is a more likely explanation than inheritance. In fact, relative to the other major maladies of our time, cardiovascular disease, diabetes, asthma, and depression, cancer is the least attributable to a common set of genetic factors. Yet it certainly involves defective DNA. It is natural to ask, then, why would evolution tolerate our genes causing so much suffering?

The short answer is that it really doesn't. There just aren't a lot of differences mulling around in the gene pool that cause cancer. Rather, cancer is a collection of diseases that all have some fundamental things in common—most basically, cells grow out of control. Control of the growth and division of cells is so complex, that inevitably it goes wrong in most individuals at some point in their life.

Similarly, for most of us, regulating our personal finances is so complex that inevitably some combination of debt, kids, and taxes puts us under financial stress that threatens to get out of control, and sometimes does. Certainly there are intrinsic reasons why some people are more susceptible (either to cancer or bankruptcy) than others, but for the most part it is things breaking in our own lives that cause the loss of control. In the case of cancer, the breaking of genes is the primary problem, more so than inheriting broken genes from our parents.

So, why should our genes break so often? The short answer here is that they really don't: It is the sheer scale of the thing that matters. There are millions of cells in each of our bodies, and every time one cell divides, just a handful of mistakes are made in the copying of the DNA. Humans should be so lucky to design a machine that makes one mistake in every billion operations. It only takes a couple of mistakes, namely mutations that change one letter of the DNA code into another, to push a cell along the winding and inevitable road to cancer.

It is somewhat pointless then to talk about eradicating, or even preventing, cancer. All we can do is contain it, and hopefully fix it before it does too much damage in each individual case. Containing it includes minimizing the number of mistakes made, for example, by reducing exposure to carcinogens such as tobacco and ultraviolet radiation. But it also means ensuring that adequate health care and timely screening are provided so that active interventions can be pursued.

In fact, probably the greatest risk factor for cancer, and the one we can choose as a society to do most about, has nothing to do with the underlying biology, but rather concerns socioeconomic status. Cancer survival is directly related to quality health care as well as the decision to take advantage of that health care, for example, by pursuing early breast and prostate cancer screening options. Without adequate health insurance this does not happen. Without a certain level of trust and confidence in the medical profession, this does not happen, and in this respect race is also a factor. There might well be racial differences in predisposition to certain cancers, but if there are, the impact of the genetics is likely to pale in comparison to the differences in utilization of the health care.

epidemiology and relative risk

One of the great heroes in the history of cancer research and public health is the little-known English physician Janet Lane-Claypon. In the first quarter of the 1900s, having become one of the first women to receive both M.D. and Ph.D. degrees, she carried out a remarkable series of studies that went a long way toward founding the field of epidemiology. One of these demonstrated the nutritional superiority of human breast milk over cow's milk and led to her being commissioned by the British Ministry of Health to conduct the first large case-control study on the causation of cancer.

After surveying 500 cancer patients and 500 matched cancer-free women she was able to conclude, presciently, that such factors as age at menopause, age at first pregnancy, number of children, and breast-feeding have a significant impact on the incidence of breast cancer. Furthermore, she was also the first to demonstrate that early surgical intervention significantly improves long-term survival. All this in half a career, since marriage in her late 40s to a colleague in the Ministry meant, by bureaucratic rule, that she must abandon her position. Such was the fate of all too many women scientists in the middle of the last century.

The epidemiological risk factors identified by Janet Lane-Claypon are now thought to reflect lifetime exposure to estrogen and other hormones involved in female reproduction. With better nutrition, girls are reaching puberty earlier, and breast maturation commences at a younger age. Delayed first pregnancy and failure to breast feed both increase estrogen levels in the breast, as does prolonged time to menopause. Similarly, prolonged post-menopausal hormone replacement therapy is now widely thought to contribute as much as 20 percent of the risk of breast cancer in older women.

Risk is a difficult concept to get our heads around. Typically it refers to the extra fraction of people who have the disease because of some risk factor. It is formulated as the ratio of the prevalence in the at-risk group to the prevalence in the population at large. A ratio of 2 means that the risk is increased 100 percent and a ratio of 1.01 means that it has increased one percent. It is far from trivial to establish that a relative risk of one percent is significant. Suppose that there really aren't any factors that make anyone in your town or subdivision or city block more or less likely to get cancer. Over a 20-year period, perhaps 500 of the 10,000 people in the neighborhood are stricken with the disease. If you arbitrarily split the community into two groups of 5,000 people each—say those who live in even-numbered houses, "the evens," and those in the odd-numbered houses, "the odds"—then you would expect that 250 each of "evens" and "odds" would have cancer. However, it would not be particularly surprising to find 230 "even" and 270 "odd" patients, just because of random sampling. In this hypothetical case, there is more than a five percent difference in the frequency of cancer in the "evens" and "odds." So, on the face of it, there is a relative risk that large of living in an odd-numbered dwelling and getting cancer.

Then, if you look at enough behavioral patterns, maybe you will discover that orange juice consumption or attendance at the opera is slightly less common in the cancer group than you would expect given the frequency of these activities in the whole community. After the statistical epidemiologists crunch the numbers, there might well be a significant correlation between not drinking orange juice or not enjoying high culture, and contracting cancer. In fact, it is all but inevitable that correlations like this arise by chance, usually at pretty marginal significance levels, but they translate into fairly large apparent risk factors. The only way around this is to replicate the study in different populations, and then to try to work out the mechanism that causes the correlation.

Of course, there is no guarantee that a strong correlation really implies causation in any case. If you look at a map of incidence of all types of cancer in America, painting states with low incidence red and relatively high incidence blue, it is virtually indistinguishable from the political map of George W. Bush's America. Voting Republican causes a lot of things, but I'm pretty sure that offering protection from cancer is not one of them.

brakes, accelerators, and mechanics

To see why cancer is a disease of the genes but not generally a hereditary disease, we need to understand how defective genes cause cancer and what makes them defective. Cancer is basically a collection of diseases, since the things that go wrong in brain tumors and breast tumors, and in lymphoma and prostate cancer, are quite different. What these diseases have in common is that the division and migration of cells gets out of control.

Basically three types of events can happen in the earliest stages of turning normal cells into cancerous ones: The brakes can fail, the accelerator pedal can get stuck to the floor, and the mechanics can go out of business. When a combination of these things happens, all sorts of other problems arise, and a cell mass that may have been relatively benign, just growing into a lump where it first appeared, starts to metastasize. This means that some of the cancer cells begin to migrate to other parts of the body where they set up secondary tumors, greatly exacerbating the problems.

Framing this discussion of cancer biology in terms of brakes, accelerators, and repairs should help make it clear that the term "cancer gene" is something of a misnomer. It implies that there are genes for cancer, when the reality is that these genes are actually essential for growth and development but can cause cancer when something goes desperately wrong. At least one percent of our genes, and by some estimates more than ten percent, are primarily involved in controlling the timing and mechanics of cell division. This is literally hundreds if not thousands of genes.

Why so complex? Think for a second about what happens after conception. You start out life as a single cell and in the space of a few weeks become billions of cells in carefully orchestrated organs with distinct numbers of retinal, muscle, neuronal, skin, and blood cells. It should be obvious that there is plenty of opportunity for things to go wrong. The more so since the development of an animal is totally self-regulating: There is no template or design that a builder can refer to. Rather, the organism unfolds with only a billion years of experience as a guide to what works.

Coordination of the parts is achieved by thousands of different signals telling each cell what its neighbors are doing and what is going on in the rest of the body. Cells can divide only after the DNA has been replicated and scanned for errors, and when it is clear that various checkpoints have been cleared. A core set of twenty or so genes orchestrate cell division in all animals (and most of these are the same in plants as well), but more than ten times this number of genes provide the checks and balances that keep the process under control.

The brakes in this process are technically known as *tumor suppressors*. These are genes whose normal job is to stop one cell from dividing into two cells, either because the body is not ready for more cells, or because all is not right inside the mother cell that is getting ready to divide. When one of these genes is mutated so that a functional protein is no longer made inside the cell, it is as if the brake lining is gone, so cell division just keeps rolling along.

The most famous of the tumor suppressors are p53 and Rb. Most if not all tumor suppressor mutations are recessive, meaning that both copies of the gene must be nonfunctional for cancer to develop. Simply put, one copy is enough, so you can think of the second copy as a backup. In fact, most of our genes are like this. *Rb* provides an apparent exception, because in half of all cases of the eye cancer retinoblastoma, which affects about one in four million children and is responsible for three percent of all cancers up to the age of 15, the cancer seems to be hereditary. Worse than that, familial retinoblastoma is often transmitted in a dominant manner, and when it is observed, most often both eyes are affected by independent tumors. The major breast cancer susceptibility genes, *BRCA1* and *BRCA2*, are like this as well.

The famous oncologist Alfred Knudson came up with an explanation for this phenomenon in 1971. Knudson's hypothesis has since become the centerpiece of a general theory for the increased incidence of cancer with age. It supposes that it takes at least two "hits" for a cell to start to become cancerous. One mutation is not enough; both copies must be knocked out, and this is exceedingly unlikely to happen early in life. However, if an individual is born with one mutant copy in all of their cells, then it only takes a single second mutation for a cancer to arise. It is apparently almost inevitable that a second mutation occurs in children who inherit one bad copy of the *Rb* gene. When this happens, a tumor starts to develop in the eye, and if untreated by laser or some other type of microsurgery, leads to blindness or can spread to the brain.

By contrast, people who have unilateral, nonfamilial retinoblastoma are unlucky enough to get two mutations affecting Rb in their own lifetime. Tens of other tumor suppressor genes can suffer the same fate. It is even possible that two mutations in different tumor suppressors combine to initiate a cancer. As we get older, the chance of two hits occurring in one of our hundreds of millions of cells increases, and so does the likelihood of getting cancer. There is precious little we can do about this.

On the other side of the axle are the accelerators. These are genes whose normal role is to push cells through their natural cycle of growth and division. Technically known as proto-oncogenes, they typically pick up information from outside of the cell in the form of growth factors and send the signal into the part of the cell where other gene products orchestrate the replication of DNA and rearrangement of the furniture that is necessary for cells to divide.

Like the tumor suppressors, these genes are essential for life. The term *proto-oncogene* means that these genes are primed to become cancer-causing genes, or *oncogenes*. Oncogenes were first identified in chicken and mouse viruses, giving rise to an early theory that cancer is often caused by viruses. This is true in some circumstances, most notably cervical cancer, but more generally it turns out that the viruses were laboratory artifacts that had picked up activated forms of normal genes with names such as *Ras*, *Src*, and *Jnk*.

Oncogenes alone are capable of promoting cancer, but it would not be accurate to say they act alone. They are so crucial that a lot of redundancy is built into the system, so losing both copies of one of the genes can actually be tolerated quite well. We can engineer mice to have activated oncogenes in all of their cells, and such mice develop tumors predictably, but only a relatively small number of cells become cancerous. So other "hits" must be required here as well. Nevertheless, there is a key difference between oncogenes and tumor suppressors: The mutations that cause a proto-oncogene to become an oncogene are ones that, instead of destroying the gene, make it active all the time.

The fiendish protein products no longer respond to signals from outside the cell; they just do their thing anyway. It is as if the cruise control gets stuck in the on position, or the accelerator pedal is stuck to the floor. There are many more ways to break a gene than to activate it like this, so thankfully, activating mutations are rare enough that only a fraction of us get cancer. The other good thing about oncogenes is that they often result in an abnormally shaped protein that can be targeted by a very selective drug that stops the protein from being active.

The third wheel of cancer genetics is the team of mechanics that repair DNA. If the DNA is broken, then so too eventually will be the instructions for making the brakes and accelerators, and many other things can also go wrong. When the major gene for familial colon cancer was identified a few years ago, *DCC1* (*Deleted in Colon Cancer 1*) turned out to encode a DNA repair enzyme.

Throughout your life, bad things happen to your DNA. Ultraviolet radiation in sunlight causes pairs of adjacent Ts to link together, and if these are not fixed, the code will change next time the cell divides. Similarly, all sorts of toxins that we eat, chemicals in tobacco smoke, and other carcinogens get into the nooks and crannies of the double helix and break it or otherwise chemically modify it. Having broken repair enzymes is not a good thing. Why the colon and breast should be particularly susceptible to this kind of problem is anyone's guess, but in all likelihood DNA repair is eventually damaged in most advanced stage cancers. Unfortunately, no drugs can take the place of the broken enzymes, so cancer biologists have to try to alleviate the consequences of *DCC* and *BRCA*, rather than fix the root problem.

I do not want to leave the impression that mutation of any of these three major components is sufficient for cancer. Current thinking is that such mutations are necessary to start a long process of tumor progression. Slowly at first, but gradually building and eventually snowballing, a mature tumor bears little resemblance to the normal cells in our bodies, as it comes to operate on its own terms. By the time a tumor is actually observed, it is vastly different from the initial cell that managed to escape the shackles imposed on it by the rest of the body. Large chunks of the chromosomes may be lost or duplicated, and tens or hundreds of genes might have picked up mutations.

An intense classical Darwinian struggle goes on inside the tumor. Any new mutation that improves the rate at which the cancer cells propagate is likely to result in that cell outgrowing the others. If you think about it, the nucleus of a cancer cell is the ultimate in selfish DNA. That one cell that starts a tumor may in the course of ten years have more descendants than the first germ cells that founded modern humans 100,000 years ago. In the language of Richard Dawkins, tumor cells are the ultimate replicators. If the sole purpose of existence were to reproduce, you'd be well advised to become a cancer cell.

However, the strategy is ultimately self-defeating. Organisms have evolved complex mechanisms to suppress cancer cells because short-term reproduction is not in the best interests of long-term survival. The community of cells is a nonselfish one in which the various parts work together in harmony. This is a metaphor for life that I find far more enticing and accurate than the nature red in tooth and claw metaphor that so often characterizes popular portrayal of evolution.

familial breast cancer

Only about ten percent of breast cancer runs in families, in the sense that sisters and daughters of an affected woman have elevated risk compared with the general population. Of this fraction, one-fifth can be attributed to two genes, BRCA1 and BRCA2. Women who inherit one bad copy of either of these genes have a lifetime risk of ovarian or breast cancer exceeding 85 percent: Each gene thus accounts for about 1 in every 100 cases of the disease. That is worth worrying about if you are in an affected family, but the flip side of this is that 97 percent of breast cancer has no known genetic basis.

Here and there, studies have by now implicated mutations in at least ten different genes in promoting specific cases of familial breast cancer. Literally thousands of known mutations are in these genes, but a handful of relatively common ones account for most of the disease that runs in families. For example, a not particularly uncommon deletion of a single nucleotide in CHEK2 is pernicious but not as bad as the BRCAs, only increasing risk perhaps twofold. If you carry this mutation, the odds are still pretty good that you'll be all right. The other mutations are so rare that they are found in only a handful of individuals. Many of the proteins that these genes encode work together as a mechanical complex that prevents something called *genome instability*. If the cell's mechanics don't work appropriately, then after a while the chromosomes basically either fall apart or divide inappropriately, leading either to abnormal growth or cancer.

Our best explanation for the relatively high incidence of these breast cancer susceptibility alleles in the human population, somewhere around two percent of all people, is genetic drift. Just like every other gene in the genome, upwards of 1 in 100,000 people are born with a new mutation in BRCA1 or BRCA2 or any of the other yet-to-be-identified tumor suppressors in the group. These individuals have the same predisposition to cancer as people who inherit a mutation from their parents instead of contracting their own. Given the late onset of disease, there is no reason why most of these people won't have just as large families and normal lives as everyone else. Consequently, the mutations can hang around in the gene pool without doing too much harm, and some of them can drift to more common frequencies.

Nature does not eliminate all of them, because it has much more severe problems to deal with, ones that act earlier in life and negatively impact more lives. Until recent times most breast cancer susceptibility mutations had a negligible impact on public health. Now that women live longer and have greater lifetime estrogen exposure, the mutations are contributing to the increased frequency of breast cancer. As small groups of people settled new lands, if a few of the founding men or women carried a susceptibility allele, it would easily have reached the kinds of frequencies that we see in, for example, Ashkenazi Jews for *BRCA1*. This is just a normal consequence of population genetics, not at all a unique property of cancer genes or Jewish habits.

growth factors and the risk to populations

Turning now to the vast majority of cases where breast cancer does not run in families, we can still ask whether genes might be involved. Groups all over the world have been tackling this question from different angles for a dozen years. For the most part they have been taking educated guesses and following up suggestive leads. In 2005, a consortium of 20 of these groups decided to pool their resources and look to see whether there were any consistent results in their combined set of more than 30,000 cases of breast cancer, admittedly mostly in Caucasian women.

The outcome was somewhat humbling. Just two genes out of the nine best guesses showed a consistent association with risk of cancer. The most convincing of these is a variant of the gene *CASP8*, which provides a modest protection against breast cancer for about a quarter of all women. The gene is involved in getting cells to commit suicide when damage to the DNA is sensed. It is the only case we know of to date where the protective variant is the less common one, where most women are at greater risk of cancer because they have the more common flavor of genetic variation.

The other gene encodes an important growth factor known as *TGFβ*, short for *Transforming Growth Factor Beta*. This gene is required for breast development in the first place. It is thought on the one hand to counteract the early development of tumors, but on the other to promote their aggressive growth after they are established. In this case, the "bad" version of the gene increases the risk of breast cancer by around seven percent in carriers and 17 percent in women with two bad copies. It is the less common allele in the population, but not by much, and about half of all women carry the risk factor.

About the same time, a group based in Cambridge, England, took a different approach. They decided to look at a lot more genes in their cohort of 4,400 patients and 4,400 age-matched cancer-free women. No single gene emerged as a significant risk factor. However, they did find a trend implicating two sets of genes as if they were gangs of teenagers looking for trouble in the neighborhood. It is tough to pin the crime on any one, but collectively they carry a smoking gun.

One group is involved in regulation of when and how often cells divide. It is not hard to imagine variation in such genes contributing somehow to cancer risk. The other group is involved in steroid hormone metabolism. This is interesting, because the leader of the gang is *ESR1*, the estrogen receptor. Recall that the combination of earlier *menarche* (the first menstrual period) and later menopause, both leading to altered estrogen levels in modern women, is now thought to be the leading candidate for the increased incidence of disease over the past several decades. If there are genetic variants that reduce estrogen exposure, it stands to reason that these would tend to be protective against breast cancer.

Subsequently, by 2007, the large consortium had a handle on another half dozen candidates after scanning the whole genome. Several of these don't immediately make a lot of sense (one is a heck of a long way from any known genes), but three of them also implicate variation in the transfer of hormonal signals inside cells.

Much the strongest case is for *FGFR2*, a receptor for Fibroblast Growth Factor, which as the name implies is a protein that normally encourages skin cells to grow. In some breast cancers, *FGFR2* is amplified in the genome, and there is evidence that different shapes of the protein produced by splicing together different bits of the gene, vary in their effectiveness. It now looks as though two-fifths of all women have some changes in a part of the gene that affects how it is expressed— either how much is made or how it is stitched together—and that these women are more likely to get breast cancer than the majority of women. **Of all the women who will have breast cancer at some point in their life, almost one half will carry the "bad" allele of *FGFR2*,** and for one quarter both alleles will be of the bad variety.

A large American study has confirmed that this is the case also for postmenopausal women who show no family history of the disease. This one gene, then, accounts for as much as one-seventh of the total genetic risk for breast cancer in the general population. That sounds like a lot, but keep in mind that most of the disease is sporadic, and not caused by genetic variation. In other words, there is no reason for anyone to rush out and get tested. *FGFR2* is simply not predictive on its own, and the great majority of carriers will never get breast cancer. In fact, putting together all the evidence from the half dozen new genes explains only another five percent, in addition to the ten percent for the *BRCA* genes, of the increased risk even in familial breast cancer.

It is highly unlikely that any more genes remain to be discovered that have effects of the same magnitude. It is much more likely that hundreds of susceptibility factors are sitting around in the genome, each with extremely modest effects or only to be found in relatively few people. Every woman probably has some of them, but it is the total constellation that elevates or depresses her risk of cancer—well, that plus lifestyle factors and raw chance.

pharmacogenetics and breast cancer

Many physicians would argue that we ought to be more interested in finding the genes that mediate how a person responds to cancer therapy than in finding the genes that cause cancer in the first place. Traditional cancer treatments were like taking a sledgehammer to dividing cells. Radiation breaks the DNA into fragments that the cells have a hard time putting back together as they are dividing, while the older chemotherapeutics generally inhibit the replication of DNA. These treatments cause hair loss, nausea, and other side effects because certain types of cells in the body are always dividing as a normal part of life, and these are affected by the nonspecific cancer treatments.

The new drugs by contrast are designed to target just the cancer cells. They do things such as interfering with the receptor proteins on the cell surface, messing up the communication pathways specifically inside cancer cells, enticing them to commit suicide, or disrupting the blood flow to growing tumors. In many cases, it is not exactly clear how the drugs work, and some really effective drugs fail in clinical application because of "off-target" effects elsewhere in the body.

Just how far genomic medicine has and has not come in the few years since the completion of the first draft human genome sequence is shown by title of an Act introduced to the U.S. Senate by Barack Obama. The Genomics and Personalized Medicine Act of 2006 (S-3822) proposed several initiatives, and a large amount of money, to accelerate the translation of genome science into clinical practices. It is foreseen that these will eventually be simultaneously tailored to the unique genetics of each patient and sensitive to the racial and environmental circumstances of the individual.

Upwards of 100,000 hospital patients lose their lives each year as a consequence of adverse drug responses. For example, 85 percent of the cases of childhood acute lymphoblastic leukemia can be treated with the drug 6-MP. Unfortunately, one in ten children carries a variant form of the *TMPT* gene, and as a consequence they are unable to metabolize the drug. If physicians know this, they can reduce the dosage and eliminate the adverse response, helping thousands of children a year. Who wouldn't want to see the widespread application of such personalized medicine?

Contemporary cancer therapy is already tailored to molecular attributes of biopsy samples. FGFR2 is just one of the many receptors that provide hints about what the cancer cells are responding to inside a person's body. Two of the other most important biomarkers are HER2 and ER, components of the receptors for Epidermal Growth Factor and estrogen, respectively.

A quarter of all advanced breast cancers make too much HER2, and this is associated with increased recurrence rates and hence mortality. In one of the early success stories for medical biotechnology, Genentech set out to specifically inhibit HER2 activity by making a molecule that binds to and inactivates the receptor. The drug, trastuzumab, trademarked as Herceptin, is actually a modified antibody, just like the antibodies that your body normally uses to fight infections.

A course of treatment costs in excess of $70,000, so health care providers have been reluctant to approve its use for treatment of early stage cancers. Increasing evidence suggests though that early intervention can be highly effective, potentially saving not just lives, but hundreds of millions of dollars a year in care for terminally ill patients. For this reason, the pharmaceutical industry is extremely active in developing new inhibitors of HER2 and other receptors like it that are implicated in numerous cancers. GSK's Tykerb, AstraZeneca's Zactima,

Novartis's Gleevec, and Genentech's Tarceva are just a few examples of such so-called tyrosine kinase inhibitors to watch for over the coming decade.

The chemotherapeutic drug of choice for combating estrogen responsive cancers has long been tamoxifen. This compound was actually first developed in the 1960s as a potential contraceptive pill. Effective for that purpose in rats, it turned out to stimulate ovulation in human women, not exactly a desirable property of a contraceptive. Further studies revealed it to be an effective antagonist of estrogen in breast tissue, and consequently an excellent drug for inhibiting the ability of the hormone to stimulate growth of Estrogen Receptor positive cancer cells.

A new generation of Selective Estrogen Receptor Modifiers (SERMs) is being introduced that get around some of the problems with tamoxifen, which is now also known to increase the likelihood of development of uterine and endometrial cancer. Eli Lilly has developed a similar drug known as raloxifene and marketed as Evista, which appears to be as effective in reducing recurrence of breast cancer, without the side effects. This drug is also unaffected by a common enzyme type, CYP2D6, that digests tamoxifen and reduces its effectiveness for some patients.

Down the road, genomics experts see a day when profiling cancers with a new technology known as *microarray analysis* will allow physicians to tailor particular drug regimens according to the entire molecular signature of the cancer. The idea is that the profile of hundreds of genes is likely to be more predictive than just the two or three that are currently examined. Hundreds of millions of dollars are being invested in this possibility, but it remains to be seen whether the technology will deliver on its promise. Currently the approach has limited approval for use with low-grade cancers where treatments are ever improving anyway.

This new approach also holds promise for guiding supplementary treatments when cancers evolve resistance to drugs such as Herceptin. Such is the competition among cells that when humans try to conquer their uncontrolled growth with drugs, the cells escalate the arms race by accumulating mutations that thumb their noses at the treatment. There are as many ways this can happen, as there are signaling pathways inside cells. Unlike normal cells, cancerous ones don't care to behave as they should; they just want to survive and divide. By looking at all the genes at once, clinicians hope to be able to target just those processes that have gone particularly sour.

why do genes give us cancer?

Why hasn't natural selection ensured that the protective versions of all the genes associated with cancer development or progression are the predominant type in the human population? The answer to this question is quite possibly a good example of the disequilibrium between our modern genetic makeup and what might have been the ideal human genetic condition throughout our history as a species.

Female reproduction is one of the traits that evolved most rapidly in humans relative to other primates. Hormonally regulated processes such as the timing and cycling of menstruation and preparation for breast feeding changed greatly a few hundred thousand years ago. This almost certainly involved selection on genes involved in hormone production over a period of thousands of generations.

Advocates of fundamentalist Darwinian medicine would probably make the argument that breast cancer is better regarded as an example of genomic conflict. They would argue that since all organisms are attempting to maximize the number of offspring they have, mutations that cause estrogen to be produced earlier in girls will tend to advance menarche and hence lead to earlier pregnancy, increasing the number of children they have. However, since as adults they are more likely to have breast cancer, there will be an opposing force of negative selection, setting up a trade-off that leads to the balance that ensures that puberty comes in the midteens.

What is wrong with this type of argument? Let's start by recognizing that earlier childbirth does not necessarily translate into having more children over a lifetime, and even if it does, it does not necessarily mean that the children will be more "fit." Birth weight is a vital indicator of child mortality and health, and is a function of maternal health. Menarche generally occurs only after a girl reaches a total body fat level of 17 percent, and the regularity of menses is also a function of growth and nutrition. Nobody knows what the relationship between early motherhood and long-term fitness may have been during human evolution.

Next we must recognize that breast cancer is predominantly a postmenopausal disease, meaning that it affects women after childbearing age and hence is not selected against with respect to having children. True, it is not advantageous for a child to have her mother die young. We also know that given the fullness of time, nature can sift through genetic differences that have an impact of just a fraction of a percent on child-bearing. So I am not saying that there is no selection against alleles that promote cancer late in life. But I am saying that it takes a lot more careful empirical observation and mathematical reasoning to establish the argument than just to make it casually.

Was the incidence of breast cancer high enough to trade-off against any possible benefits of early pregnancy? It is almost impossible to know and seems unlikely. Any trade-off argument is a gross oversimplification. Estrogen regulates hundreds of coordinated processes; timing of menarche and menopause are changing against a background of dramatic remodeling of reproductive strategies and of primate lifespan, not to mention mortality and health risks. A few hundred thousand years is a very short time to expect the genome to come to some sort of equilibrium.

What we can be confident about is that there is genetic variation affecting estrogen production, that this variation has been under selection throughout human history, and that any connection to breast cancer liability is likely to be incidental. Now throw in all the changes that modern society has wrought—excellent childhood nutrition, cultural taboos against teenage pregnancy, the obesity epidemic, stresses on nuclear families and social relationships, longevity—and whatever effect that variation had in the past is turned upside down. Don't expect a new equilibrium any time soon, and don't expect the genetic risk factors to go away either.

A couple of naïve but legitimate questions remain that deserve to be asked about why cancer is so prevalent, accounting for as much as a fifth of human mortality, if this natural selection thing is supposed to be so efficient. Why would there be cancer genes at all, and why hasn't the evolutionary process done something about them? I do not have a quick sound-bite answer, other than to reiterate that cell division is so complex that there is ample scope for it to go wrong, and that the high incidence of cancer is really a modern phenomenon. Let's recap the salient facts from this chapter that might help us to understand this better.

First, the terminology "cancer gene" is at best misleading and definitely inappropriate. It implies that there are for some reason genes whose job it is to cause cancer, just like there are viruses and bacteria that seem to exist solely to cause misery. The reality is that hundreds of genes that are perfectly good and vital citizens of the genome unfortunately mutate into forms that contribute to cancer. The most insidious mutations affect genes whose function it is to protect the genome by repairing the DNA: When these stop working, the genome starts to fall apart, and cells lose control of their place in the organism.

Every single human carries mutations in several of these genes, but since a lot of redundancy is built into the control of cell division, it does not much matter. At least, not until new mutations build up in our cells during the course of life, knocking out these failsafe mechanisms. This is why chance plays the major role in determining who will get cancer: It is just a stochastic matter of who is unlucky enough to find themselves with a bad combination of mutations that the body cannot eliminate.

Second, we shouldn't blame the genes for cancer: By far the greatest threat comes from environmental factors. Smoking or hanging out in smoky bars, not eating your greens, tanning in the midday sun, and the combination of early puberty with delayed pregnancy, are all to blame. These are things we can do something about, though reality usually steps in the way, and most of us make a Faustian pact to trade a little extra cancer risk for the pleasures of social engagement, healthy looks, or an independent career. The one thing we cannot do anything about, though, is the biggest environmental factor of all, and that is growing old.

Cancer is fundamentally a product of disequilibrium between our genome and our culture. All of a sudden, in the space of a couple of generations, humans are living 20 years longer than ever before. Cryptic genetic susceptibility factors that have only a miniscule effect up to the age of 50, too small an effect for natural selection to do anything about, are now uncovered. These variants account for maybe a quarter of all mortality in old age, but they just have not played a significant role in human history hitherto.

Third, those cases where we can blame specific genes, account for only a small proportion even of cancer that seems to run in families. BRCA1 and BRCA2 are the two best-known susceptibility genes. If you are a woman who inherits a mutation in one of these two genes, you have a high lifetime probability of having breast cancer. Literally hundreds of mutations of this type are in the human gene pool, most very rare, but some at a frequency as high as a few percent in particular populations.

Yet, the mutations only account for one-tenth of familial breast cancer, which in turn is one-tenth of all breast cancer. Geneticists are just now uncovering a few other genes that harbor mutations that also contribute, but the bottom line is that we just do not know what genes account for the majority of breast or any other type of cancer.

In fact, we don't really have a good model for how they act either. Knudson's two-hit hypothesis has served pretty well as a starting point. The idea is that you need at least two mutations to initiate cancer. If you inherit one from your parents, you are off to a bad start, because you only need one more during your life. Since that first mutation is insufficient to cause cancer, it can hang around in the gene pool, ever so slightly increasing the incidence of cancer and contributing to its tendency to run in families. Presumably there are hundreds of genes harboring such mutations and hundreds of different mutations in each of these genes, each of which increases cancer susceptibility a fraction of a percent. Eventually we will find many if not most of these, but it is doubtful it will do us much good.

Though not discussed here, a lot of cancer susceptibility has more to do with what happens in the progression from initial lesion to mature tumor, than in the initial appearance of cancer. The potential of cancer cells to *metastasize* (start migrating around the body), their ability to *vascularize* (attract blood vessels to support malignancy), and the capacity of the immune system to detect and deal with an early stage cancer are all affected by our genes. Our psychological and nutritional status can play an enormous role here as well.

Putting all this together, the inescapable conclusion is that we are dealing with an immensely complex foe. The media, naturally, likes to report from time to time that scientists have identified a new gene for cancer. Our response is to assume that this is something like a gene for blue eyes or male-pattern baldness: If you get it from your parents, you are going to get cancer, and the reason why we cannot yet predict cancer is because we haven't found the genes yet. The truth is otherwise.

3

Not so thrifty diabetes genes

jackie and ella   Up to a third of the world's population may be diabetic by the next century, and it can afflict just about anyone.

the pathology of diabetes   There are two major types of diabetes, both due to lost regulation of blood glucose by insulin.

type 1 diabetes   The rare form of diabetes arises because a child's own body destroys the pancreatic cells that make insulin.

an epidemic genetic disease   Fast foods and sedentary lifestyles add up to a whole lot of extra pounds and burgeoning diabetes risk.

genetics of obesity   Hundreds of genes contribute to whether a person is likely to put on weight more than average.

type 2 diabetes   Several alleles have recently been identified that promote diabetes in Caucasians, but they are slowly being displaced by protective variants in the human gene pool.

debunking the thrifty genes hypothesis   Arguments that diabetes is caused by alleles that favored rapid assimilation of carbohydrates in times of famine fall short.

disequilibrium and metabolic syndrome   We've pushed our genetic legacy to the limits of its ability to cope with modern diets and stress.

jackie and ella

Diabetes is the global warming of public health. A crisis that will affect the lives of literally billions of people is looming with an inevitability that borders on irrepressible. We pretty much know the causes and even have a fair idea about how to prevent it. Yet inertia is such that somewhere between one-quarter and one-third of the world's population will become diabetic in the next few generations. These are the proportions about to beset North America and Europe, but the developing world is rapidly following in step. So pervasive is the switch to fast foods and sedentary lifestyles that obesity and its consequences will soon replace malnutrition as the predominant food-related malady of even the poorest nations.

Diabetes is also an equal opportunity killer. Young or old, man or woman, rich or poor, black or white, thin or fat: Everyone is at some risk. The list of prominent diabetics cuts across professions and includes some surprises that buck the popular impression that it is solely a disease of the morbidly over weight. Spencer Tracy, Mary Tyler Moore, and Halle Berry head a long list of actors; Menachem Begin, Anwar Sadat, and Mikhail Gorbachev suffered; Howard Hughes was afflicted as was, ironically enough, Ray Kroc, the founder of McDonald's.

Like tens of millions of ordinary folks, these individuals were able to overcome daily discomfort and pain to lead highly productive lives. In most cases, though, the disease exerts its authority in the end. The brief life stories of two of the most influential African Americans of the twentieth century tell how.

Jackie Robinson was the man who broke the color barrier in professional sports. Born in rural Georgia and raised by a single mother in Pasadena, California, he encountered plenty of prejudice as a kid, yet managed to turn sorrowful lessons into his positive expression that "there's not an American in this country free until every one of us is free." Despite lettering in four sports, financial difficulties forced him to leave UCLA early and join the Army. That career was cut short by a court-martial and subsequent honorable discharge following his objections to racial slights such as being ordered to the back of the bus—11 years before Rosa Parks famously refused to do so. After one season in the Negro Baseball League, Jackie was noticed by the president of the Brooklyn Dodgers and invited to the big leagues. He went 0 for 4 in his first game on Opening Day, April 15, 1947, and took 21 at bats over a week to get his first hit, but then went on to win Rookie of the Year honors, a Most Valuable Player Award a couple of years later, and eventually induction to the Hall of Fame. A lifetime .311 hitter, extraordinary base runner, and natural second baseman over a ten-year career, Jackie Robinson would have been one of the greats of his era even had he not also changed the face of the game.

Tragically, Jackie's life was cut short by heart disease at the age of 53, just 15 years after the end of his playing days. His relatively short life had been dedicated to opening doors for others, as a campaigner for the NAACP, and as a businessman, notably founding a construction company that built housing for low-income families. In the dozen years after retiring from baseball, he had gone almost blind, while chronic heart problems had taken away his athleticism. We could be forgiven for assuming that it was the burden of bearing daily insults while carrying the torch for a new generation of athletes that had eaten away his strength.

In actuality, it was the burden of too much glucose in the blood, slowly but surely eating away at the cells of his retina and his heart muscle. No doubt, stress did not help. The combination of high blood pressure with depositions of fatty materials inside the artery walls increases the demands on the heart. **Diabetics who have already had one heart attack, as Jackie Robinson had a few years before his death, are at extreme risk.** He didn't have the benefit of our knowledge about the effects of saturated fats and of smoking, nor did he have access to all the modern drugs that make diabetes manageable.

Ella Fitzgerald's story is equally inspiring, her battle with diabetes equally devastating. Someone once said of Ella that, "Music comes out of her. When she walks down the street, she leaves notes." In a career that spanned five decades, she became synonymous with scat, legendary for the clarity and range of a distinctive voice that often borrowed from the horns in the big bands she accompanied. Orphaned at the age of 15, she managed to survive the endemic abuse of a reform school for girls, only to be turned out onto the streets of Harlem. In 1932 she was discovered at an amateur night and quickly found herself the leading lady of jazz, performing to packed houses at the fabled Savoy Ballroom, collaborating with all the greats: Count Basie, Duke Ellington, Nat King Cole, and Charlie Parker. A shy girl, born to poverty and disadvantage, she turned an untrained voice into one of the most improvisational instruments of a golden era, and one of the surest cures for the blues ever devised by humankind.

Inexorably, though, diabetes took over. Her eyesight fading, circulatory problems started to eat away at her body, and at the age of 66 she underwent quintuple coronary bypass surgery as well as heart valve replacement. Eight years later, ongoing heart disease drained the vessels of her legs of the lifeblood they needed, causing her to lose both to amputation. Then in 1996, the 81-year-old Ella Fitzgerald's heart finally gave way due to complications of diabetes. Today, her Charitable Foundation, like Jackie Robinson's, carries on her legacy. It provides educational opportunities for children; fosters a love of music; provides health care, food, and shelter for the needy; and supports medical research relating to the eye and heart disease caused by diabetes.

the pathology of diabetes

Diabetes mellitus is actually now recognized as a spectrum of diseases that converge on a common symptom, chronically high blood glucose. The name translates roughly from the Greek and Latin into "passing through honey," meaning that a diabetic's urine has a sweet taste. I'm not sure how the English physician Thomas Willis worked this out. Maybe urine tasting was more common back in the seventeenth century.

For some time, physicians have recognized three forms of the disease, which we know as juvenile, adult-onset, and gestational diabetes. Since so many obese children are now contracting the adult form in their teens, these terms are out the door and have been replaced by *insulin-dependent* and *non-insulin-dependent diabetes mellitus* (*IDDM* and *NIDDM*, respectively). Since these are confusing, we will stick to type 1 and type 2, or more simply T1D and T2D. The gestational form afflicts pregnant women and can be thought of as a special class of T2D that thankfully usually disappears after childbirth.

By far the more common of the two types of diabetes is T2D. This is the epidemic form that is now seen in more than 20 million Americans and 170 million people worldwide. Another 40 million Americans are considered prediabetic and at risk for full-blown disease. T1D by contrast has a relatively constant prevalence of a fraction of a percent of the population. Very few environmental factors are even suspected to increase the likelihood that a child will have T1D. By contrast, T2D is very much a product of modern diets and modern lifestyles.

The major difference between the two forms is the way that hyperglycemia (high blood glucose) comes about. When we eat, the food is rapidly broken down into sugars, and in particular into a sugar building block called *glucose*. This is the major source of energy for the body. As anyone who has ever taken college level biochemistry knows all too painfully, most carbohydrates are broken down into glucose through convoluted pathways of metabolism. Sheep and cows just let their food ruminate in the stomach for days, slowly releasing sugars, but other mammals have a highly evolved way to regulate glucose so that the levels stay fairly constant throughout the day. This way we can handle more complex and diverse diets, dining at Spago's should we wish, instead of on the front lawn.

The main way that glucose levels are regulated is through the hormone insulin. Insulin tells your fat and muscle cells that you've just eaten and that they should be prepared to take up the sugars that find their way into the bloodstream. If insulin isn't working, then glucose metabolism is out of whack and bad things follow. In T1D, insulin doesn't work because the cells that make the hormone have been destroyed. T1D is an autoimmune disease, just like arthritis, lupus, and multiple sclerosis. The difference between these diseases lies in the nature of the cells that a person's own immune system destroys. In T1D it specifically attacks just the islet beta cells in the pancreas. Without those cells, no insulin can be made, and glucose remains in the blood. In T2D by contrast, insulin levels are fine, but the body does not respond to the hormone. It is said to be insulin-resistant. You get the message, but just don't respond to it.

It follows that treatments for these two classes of diabetes need to be different. People with type 1 diabetes must take insulin on a daily basis, mostly by injections that after a time become extremely painful. Drug companies are working on forms of the hormone that can be taken orally, but it tends to be digested before it can get to the liver where it is most needed. Long term, it will be better if physicians can work out ways to transplant healthy islet beta cells back into the pancreas, possibly using stem cell technology. Type 2 diabetics manage the disease with a variety of drugs that act principally to lower blood glucose levels. Even better, changes in lifestyle such as increased exercise, healthier eating habits, cessation of smoking, and generally reducing stress can all turn the tide of disease.

What is so bad about high blood glucose? If the whole point of eating is to provide energy for cells, what can be the problem with allowing them to bathe in sugar? **Basically it is a matter of all things being better in moderation, that sometimes too much of a good thing can be bad for you.** Harm manifests itself at several levels.

One of the most important is actually an indirect side effect of cells not absorbing glucose. Even though the glucose levels may be high in the blood, if the cells do not take up enough of the sugar, because they are not responding to insulin, they are forced to use a different source of fuel. This is the reserve of fats and proteins that they have. But burning fats produces things called *ketones*, which are acidic. If there are too many ketones, then the pH of the blood drops. If it gets below 7.35, the result is acid rain in the body: Our cells don't like acid any more than trees do.

Turning to the direct effects, often the first problem is that the kidneys are upset. Their role is to clean up toxins in the blood, which they do by filtering them from the blood into what will become urine. The whole system depends on appropriate osmotic pressure, on the balance of electrolytes and sugars. If too much sugar is in the blood, it must be pumped into the urine, and excessive urination—often the first sign of diabetes—ensues. The lost fluids have to be replenished somehow, generally by excessive drinking (preferably of water), lest the body coaxes its own cells into giving up their water. Should this dehydration happen to cells in the brain, dizziness and fainting spells will result, and in extreme cases even unconsciousness.

Eventually the imbalance puts so much strain on the kidneys that they start to fail. Diabetic nephropathy is the most common cause of the need for dialysis in the United States. Once pathology starts to develop, it can progress quickly, particularly if blood pressure is poorly controlled.

A second common sign of diabetes is blurred vision. For one thing, constant absorption of glucose changes the shape of the cells in the lens of a person's eye, which affects eyesight. But the bigger problem is that the blood cells of the retina are very sensitive and easily damaged by buildup of obstructions that cut off oxygen supply and kill the cells. When this happens new blood vessels start to grow in, and blood can seep into the back of the retina, so clouding the image that blindness ensues. Diabetic retinopathy will affect 80 percent of patients who have the disease for 15 years or more, unless they take careful steps to control the symptoms.

Local damage to the small blood vessels throughout the body has other consequences. As glucose builds up, the vessels thicken and can constrict, resulting in mini strokes in the periphery of the body rather than the brain. These disrupt the general flow of blood. Starved of oxygen, nerves eventually stop working as diabetic neuropathy sets in. Impotence may be one concern, but numbness and diarrhea, loss of bladder control, and muscle weakness are all common signs of advanced disease.

From there, things can only get worse. Completely deprived of blood flow, the arms and legs start to waste away, and deprived of feeling it is easy for cuts, bruises, and calluses to turn into ulcers and open sores if not gangrene. Diabetes is the primary cause of nontraumatic amputation. The heart itself also needs a constant supply of blood, so cardiac failure is common. As blood pressure rises, so too does the risk of atherosclerosis, particularly in obese or hypertensive patients. Ultimately, heart disease and stroke are the most deadly consequences of all forms of diabetes.

type 1 diabetes

Somewhere in the vicinity of 1 in every 300 children will contract the fat-free form of diabetes, T1D. We currently have no way of identifying those at risk in advance and are powerless to stop the progression. On the other hand, we can be fairly confident about who will not be type 1 diabetics. The reason is that almost everyone who has the disease has one of a handful of genetic markers at one particular place in the genome. The Human Genome Project is telling us that they are likely to have a few other risk factors as well.

Even a cursory look at the distribution of T1D in families makes it clear that there is a sizeable genetic component to the disease. If one identical twin has it, then just over half the time the other one will as well. By contrast, regular brothers and sisters of a type 1 diabetic have only a 1 in 20 chance of contracting T1D, reflecting the fact that they only share half of their genes. But this incidence is itself 20 times greater than the prevalence of the disease in the whole population.

Starting in the 1980s, several groups of researchers across the world set out to see whether they could find those parts of the genome shared by people with T1D. This was to be the perfect situation for finding the genes that contribute to complex disease since there is a willing group of patients and neither the environment nor behavior seems to be much involved. Their strategy was to concentrate the search in families. By now, more than a dozen regions have been linked to T1D in multiple studies, and in the past few years half of them have been associated with a particular gene.

The surprise is that after all this effort, just one complex of genes stands out and explains almost half the incidence of T1D. This complex is the central player in immunity. Not so surprising, the next biggest factor is the insulin gene itself. The way it seems to promote disease was unexpected but certainly makes sense in hindsight. The other players are a supporting cast, interchangeable and barely noticeable by themselves.

The genes in the major risk complex are called *Human Leukocyte Antigens (HLA)*. They live in a community of hundreds of genes known as the *Major Histocompatibility Complex (MHC)*. These are perfectly reasonable names for an immunologist, but we'll stick to HLA and MHC. The MHC is incredibly diverse, both in the range of things it does, and the variability you find in it. We'll meet it again in the next couple of chapters.

The HLA genes are thought to be the most variable part of the human genome. Aside from immunity, one of the other things the MHC does is help us determine whether other people are genetically different. Believe it or not, we are constantly sniffing one another out. Happily, most of us are a little more subtle about it than dogs are, but studies of young adults indicate that they prefer mates who look different with respect to the makeup of their MHC.

Think of your HLA proteins as the many different picture hanging devices you can get for mounting things on the walls of your house. You've got your basic nail, then there are hooks of various sizes and shapes that slide over a nail, or there are ones with wires, and others that are sticky or made of Velcro. Inevitably, you don't have the one you need lying around the house, but can usually make do, and if not, then a trip to the hardware store will solve the problem.

The HLA proteins basically hang little broken up pieces of molecules on the outside of your cells, where the policemen of your immune system, the T-cells, can look them over and decide whether they indicate trouble and whether to do something about it. Some viruses and microbes are able to avoid the immune system because their proteins have shapes that make them difficult for HLA to bind to. Since the range of pathogens is so great, the HLA must be diverse enough to show as many of them as possible. But evidently they cannot hang everything, so there is a constant turnover of variation based on the tug-of-war between pathogens and our collective immune systems. You'll find different flavors of the HLA in different populations, reflecting the different recent history of infectious agents.

Soon after birth, your immune system has to make a whole bunch of decisions about which of the molecular pictures hanging on the walls of your cells are unlikely to indicate harm. In other words, it has to be able to identify your own proteins and make a memory of the difference between these and foreign ones. This is called distinguishing between "self" from "non-self." When something goes awry at this step, it is likely that at some point your body will start attacking itself, with the immune system destroying your own cells as if they were bacteria or other foreign cells.

In T1D, the immune system fails to recognize insulin as self, and so as babies grow into children, it starts destroying the cells that make insulin, the islet beta cells in the pancreas. It looks very much as though two major genetic factors can contribute to this happening.

The first is if your HLA has the wrong set of picture hangers. It is as if when your mother-in-law shows up with the family portrait that must be hung over the mantelpiece where everyone in the world can see it, you just can't bring yourself to hang the picture, and suffer the consequences forever. In Caucasians, the combination known as DR3/DR4 is the worst, while in Asians it is DR4/DR9. A change of one amino acid in DR4 probably disrupts the ability of HLA to bind to insulin, so insulin is not appropriately shown to the immune system. Three percent of white folks have the DR3/DR4 combination, having inherited either of the two from each parent. They have a fifteenfold greater risk of contracting T1D than everyone else, and account for 30 percent of all type 1 diabetics. If you don't have either of these flavors, it is unlikely that you will have T1D. But if you do have the risky combination, you aren't necessarily going to be diabetic.

It stands to reason that this deficit could be overcome by producing more insulin, forcing the HLA to show insulin to the developing immune system, whether it has the right hangers or not. This is the reason why the insulin gene itself is the next major risk factor. There is a very odd little repetitive stretch of DNA right in front of the insulin gene in humans. Anyone with hundreds of copies of the stretch makes more of the insulin protein where it matters, which is not the pancreas but rather the thymus. The thymus is the police station where the immune system does its surveillance. If more insulin is made there, it is more likely that insulin will be recognized as part of "self" and tolerated in the body. Later in life, the immune system will not attack the beta cells in the pancreas.

Unfortunately, only about 1 in 5 of us has the highly repetitive form of the insulin gene. Instead, the vast majority of people are at risk for T1D on account of having two copies of the other allele, the one without the repeats that causes less insulin to be made in the thymus. A consequence of this is that even though the repetitive allele makes quite a difference for any given individual, it is not protecting enough individuals to account for a whole lot of the variation in susceptibility in the population at large.

Inadvertently, it seems that northern Europeans might have been playing with their infants' insulin levels over the last couple of generations. The baby formula derived from cow's milk that contains bovine insulin might somehow be interfering with the establishment of self-recognition of human insulin. Studies in Scandinavia and Germany have repeatedly shown that mothers who stop breast-feeding after three months elevate the risk of their child having T1D, perhaps as much as twofold. This effect has not been established in North America, possibly because formula here is processed differently.

Two other genes that have been implicated in T1D have modest effects of much less than a twofold increase in risk, and like HLA they appear to be different in different racial groups. Their names, *PTPN22* and *CTLA4*, conjure up images of everyone's favorite *Star Wars* robots, R2D2 and C3PO, and indeed both are part of an elaborate T-cell machinery. These are the cells that recognize and respond to foreign invaders and usually leave familiar molecules such as insulin alone. The details are still being worked out, but particular alleles make it less likely that T-cells that do recognize insulin are eliminated from the blood.

I cannot resist finishing this section by making reference to the fifth T1D risk gene, *SUMO4*. Believe it or not, it has only been clearly established as a risk factor in Japan. The name is pure coincidence, but you can imagine that there are plenty of references to wrestling with diabetes genes in Asia. *SUMO* stands for *Small Ubiquitin-related Modifier*, and the SUMO proteins are involved in—you guessed it—sumoylation of other proteins. Just how this relates to T1D is not yet clear.

an epidemic genetic disease

The story for type 2 diabetes is really quite different. T2D is essentially an epidemic disease. Prevalence has increased from a few percent to well more than ten percent over the past 30 years and continues to trend upward at an alarming rate. Genes alone cannot cause an epidemic; there must be some environmental agent. And we all know what that agent is: the transition to a fast food, slow couch-potato lifestyle. The genes are just accomplices—from their viewpoint unwitting ones. In the blame game, they are innocent victims of changes that humans have wrought upon themselves, caught up in a disease they have no business being associated with.

We can talk about hyperglycemia and insulin resistance as much as we want, but the root problem in T2D is that regulation of metabolism is out of control. Constantly exposed to high sugar levels in the diet, we produce insulin at higher levels than the body evolved to tolerate. Eventually it cries wolf, shutting down its response to the hormone. The modern lifestyle has pushed an exquisitely evolved system of checks and balances to the limits of its buffering capacity. Those who are unlucky enough to be genetically less buffered find themselves more susceptible to developing diabetes.

So genes are involved, but more as a facilitators than causal agents. If we want to understand what they are doing, we need to address three questions. First, why are some of us more prone to overeating than others? Second, why does overeating lead to obesity more readily in some than in others? Third, what is the relationship between weight gain and T2D, and are there genes that contribute to T2D independent of obesity?

It is not difficult to see how weight gain gets out of control. Just a small change in the ratio of caloric consumption to expenditure adds up over time. A 40-year-old man who is 20 pounds overweight, which is almost the norm these days, has been putting on an average of a pound a year since he left school. That equates to just 10 grams a week. How many grams of sugar are there in a can of Coke? 39. In theory, cut the extra beer for the road or the donuts at the weekly group meeting, and problem solved.

We all think, dipping our hand into a co-worker's candy jar, that we will work it off on the walk back to our own office. At least the Europeans get to walk to the tram instead of the garage every morning. If simply adding up the calories translates into inches around the waist, it really doesn't seem right that three half-hour workouts a week don't add up to a lot more weight loss.

Other factors must be in play here. A major one is the transition to sedentary lifestyles that occurs when most people are in their twenties, especially in the modern economy where work is more likely to involve sitting in front of a computer than being active outside. Even if most of us get the eating part of the equation more or less right, too many of us don't find time for the physical.

Another factor is socioeconomic: It is clear that obesity is proportionately a much greater problem for the less well off. In the space of a century this represents something of an inversion, since malnutrition in the developed world is now much less common than undue weight gain. The culprits are self-esteem issues and fast food. It is sadly most difficult to lose weight when you feel badly about yourself, when you notice that success goes to attractive and energetic people, and a negative cycle of worthlessness sets in. Diets never take effect straight away, and exercise programs usually make you feel really tired for the first few weeks, without producing results. It is easy to give up.

At the same time, chains of McDonald's, Taco Bell, and Bojangles beckon with promises of cheap and tasty meals, supersized for just a few cents extra. Dollar for calorie, energy dense burgers, nuggets, and sodas—all in some way or another just processed corn—are three times better value than the South Beach foods we should be eating. A few dollars for a morning sausage biscuit and coffee seems like a good deal. The cost of feeding a family of four from one of these chains is rarely more than $30, but put together a nutritionally well-balanced and freshly spiced meal from Martha Stewart's *Living*, and it will run you $50 plus. Equally important is budgeting time into lives filled with second jobs, kids' needs, and a simple desire to collapse in front of the TV.

In fact, our entire food culture is set up around rapid consumption. Michael Pollan's fascinating and frightening book *The Omnivore's Dilemma* explores the reshaping of the American food economy starting with the industrialization of corn. He gives a hefty wag of the finger to Richard Nixon's Secretary of Agriculture, Earl "Rusty" Butz, whose policies led directly to the massive surpluses of corn production that drive the feed mills of Kansas and the wet mills of the upper Midwest that process kernels into a thousand varieties of bottled glucose.

The end product of the tens of billions of dollars of government subsidies behind the oceans of maize that float across the great plains each summer is in a very real sense the thickened waistlines and the clogged arteries of the modern suburbanite. Only, those at the bottom of the socioeconomic food chain are the most heavily affected. Next time you pick up a fast food meal, just double the price mentally and put the difference toward the several hundred dollars of health care premiums you're paying to cover the billion dollars of heart disease treatment we take for granted.

Despite temptation, our genes do have a say in establishing how each of us responds to excessive caloric intake. We all know fat people who seem to eat like a kitten, and jealously regard those who can eat whatever they like yet slip into a size two comfortably. Hollywood caricatures of big fat men exploding as they shovel down yet another turkey leg aside, the morbidly obese are just as much victims of a raw hand in the genetic lottery as from their eating habits.

genetics of obesity

Anyone in any doubt about the power of genes to influence weight gain need only consider the case of obese mutant mice. These are a strain of otherwise normal mice that appeared in a laboratory colony as a result of a spontaneous mutation in 1950. Animals that inherit two copies of the mutation are really big, up to four or five times bigger than littermates with just a single bad copy of the affected gene, big enough to swallow up their siblings in the flabby folds of their skin. They get this way because they are unable to control their appetite, and just keep eating.

In the mid-1990s it was discovered that the *obese* mutation knocks out a gene that encodes the peptide hormone leptin. Leptin is one of the primary signals that the brain uses to stop us eating when it senses that we've had enough. The crucial part of the brain is called the *hypothalamus*, which among other things is also known as the satiety center. In other words, we don't just stop eating after a meal because there's no more food on the plate, but rather because we have finely tuned sensors that actively tell us it is time to stop eating. Disrupt those sensors, and we keep eating, and obesity is sure to follow.

Imagine the glee with which would-be pharmaceutical giants met this discovery. Surely administration of leptin as a drug would provide the panacea for weight loss that would slip into the void left by the tragic demise of Fen-Phen. Alas, it turns out that obese people actually tend to have higher levels of circulating leptin than people of normal weight, indicating that they have become resistant to it. In a small number of cases of morbidly obese families the leptin gene is deleted, just as in the mice, but it appears that the gene plays only a minor role, if any, in general human obesity.

Three varieties of drugs do seem to work to control weight gain: appetite suppressants, carb-blockers, and fat-burners. Fen-Phen is an appetite suppressant that, like several other drugs, acts to reduce signaling between neurons by serotonin in the hypothalamus. Not surprisingly, these drugs have many side effects, including upsetting the heart, so you should take them at your peril.

Appetite regulation gets a lot more complex the more physiologists look into it. A complex network of signals keeps everything in the appropriate balance, preventing excessive eating habits that lead either to obesity or anorexia. The range of hormones that moderate energy and glucose homeostasis reads like Santa's reindeer: go leptin and visfatin, on adiponectin and omentin, there's ghrelin and resistin, and oxytomodulin and amylin, not to mention peptide YY, glucose-dependent insulinotropic polypeptide, and the glucagon-like peptides. All these need to be integrated with daily rhythms—appetite must be suppressed while we sleep—and with how we're feeling.

The network of interactions likely remained almost unchanged over tens of millions of years of primate evolution but has suddenly had to cope with the double whammy of first a dramatic change in human body size and now a fundamental shift in diet. It is no wonder that the system is confused by the constant availability of sugary foods in the modern world, no wonder that the natural balance is upset; the genome is out of equilibrium with the modern world.

Just this short discussion has suggested 20 or so genes that may be involved in weight gain. Each one of these genes is a candidate for a place in the genome where variation could contribute to the obesity epidemic. We haven't even begun to talk about digestion, fat deposition, energy burning, or basic metabolism, and if we did the list would rapidly expand to more than 100 genes.

Studies over the years have actually implicated more than 250 places in the human genome that might lead toward obesity, without actually finding the culprit genes. Individually these studies are barely worth the paper they are written on, and most lead to investigative dead-ends. Collectively though they tell a truism that weight gain really does take a genome.

In other words, we need to completely abandon the notion that there is a gene for obesity, or even that there are a few genes for obesity. Instead embrace the concept that hundreds of genetic variants are a part of the normal regulation of body weight, and it is an inevitable corollary that some individuals have combinations that predispose them to disease.

This is the notion introduced by an analogy in Chapter 1, "The Adolescent Genome," that many times companies fail not so much because of an incompetent CEO, but rather because of the accumulation of myriad natural incompatibilities. Every company deals with employees going through a divorce, coping with rebellious teenagers, pushing their own agenda at the expense of the team, or struggling with the latest software. Change the pressures slightly, and a relatively functional group can become dysfunctional. Something like this is contributing to the obesity epidemic.

What has been discovered by randomly testing hundreds of thousands of variable places in the genome to see which ones are correlated with obesity? One striking result is that a gene called *FTO* has a lot to say about who is overweight, across most human populations. The 16 percent of adults who have two copies of a particular allele of the gene *FTO* are about one and a half times more likely to be obese than everyone else. Conversely, the 36 percent of adults who have both copies of the other allele have almost half the likelihood of being obese, and they are on average 5 pounds lighter. This conclusion is based on measuring 40,000 people in 13 different studies, so there is little doubt about it. Shockingly, the effect of the gene starts to appear as early as seven years of age, before the kids themselves can be expected to take responsibility for their eating and exercise habits. That is definitely not to say they cannot do anything about it as they get older, but the deck is stacked a little against them from birth.

Unfortunately, as yet we have almost no idea what *FTO* does, where it does it, or why the different flavors do it differently. Surely it will not take long for scientists to figure this out, but it is one of the frustrating things about contemporary genetics. Like rising gasoline prices, often we can see what the problem is but can't do anything about it.

Another example of this is the gene called *INSIG2*. My first thought on reading the paper describing the discovery of this particular obesity gene was that the researchers must have a wry sense of humor and a fearless attitude toward funding agencies. So many findings of highly significant associations between genes and diseases have turned out to be false leads that it is almost asking for trouble to give your gene a name that conjures up "insignificance." But it turns out that *INSIG2* actually stands for *Insulin-induced gene 2*. The protein encoded by the gene seems to be involved in the synthesis of fatty acids and cholesterol, which obviously makes sense if you are looking for something to do with obesity.

The actual DNA changes that alter the function of the gene have yet to be identified, but three different studies of several thousand Caucasians on both sides of the Atlantic, and of African Americans, suggest that the ten percent of individuals who have two versions of the less common allele are more likely to be obese. The risky variant is at a relatively constant frequency of around one-third of the alleles in all populations examined, but oddly it does not seem to promote weight gain in all populations. The obesity-associated allele seems to be the ancient one that was present in humans before they began moving around the globe and practicing agriculture. The protective type is the more modern one, implying that we are evolving a genetic constitution that is less predisposed to putting on weight.

Another gene, *ENPP1*, also known as *PC1*, popped out of a search for genes that might be involved in diabetes. It encodes a protein that binds to and turns down the function of the insulin receptor. If you think of the insulin receptor as the main control on your dashboard that allows you to work the air conditioning in your car, then *ENPP1* is like the knob on the air vent that provides a little more control. Cell biology is full of such devices that you can do without, but that make life easier.

A common form of ENPP1 in humans has one different amino acid, affecting how the protein binds to the insulin receptor. Many other alleles also affect how much of the protein is made in particular tissues such as the pancreas, the liver, and fat cells. It is easy to imagine that such different forms affect the development of resistance to insulin, and hence susceptibility to diabetes, and it seems that they do. But it turns out that they also predispose carriers toward obesity. Not a lot, but enough to make a difference, at least in Caucasians: One study of Japanese found no effect at all. This is emerging as a common occurrence in the genetics of disease: Whether a variant flavor of a gene matters is peculiar to each population.

Just like cancer, a few percent of morbid obesity is actually explained by heritable cases that can be attributed to single severe mutations. One of the main culprits is a gene that mediates the feeling of satiety, called the *fourth melanocortin receptor, MCR4.* Melanocortin is one of those fascinating genes that seem to be involved in everything. Variation in one receptor gene, *MCR1*, is responsible for the white patches on many of our furry friends, while another one is implicated in erectile dysfunction, and seasonal affect disorder ties in there somehow as well. Numerous studies now encompassing tens of thousands of cases establish a weak link between common variation in *MCR4* and obesity. I can't wait for the advertising campaigns that, after drugs are developed that overcome the genetic legacy here, solemnly warn that if you experience an erection lasting for more than four hours while trying to lose weight, consult your doctor.

type 2 diabetes

Diabetes is to obesity as crime is to unemployment. You can have one without the other, but the latter certainly leads to the former. Consequently, the obesity gene map heavily overlaps with the diabetes gene map, though we now know that it is not sufficient. Many common polymorphisms contribute to diabetes without playing much of a role in normal weight control.

The body has evolved exquisite mechanisms based around insulin to keep glucose concentrations in balance. Normally it copes with excess by storing it as fat. But if the system is overloaded for too long, it becomes stressed, and eventually throws in the towel. Diabetes then arises from a double dose of disequilibrium. First is the disequilibrium between genes and the environment; second is the loss of equilibrium in the balance of hormones that function to ensure that energy reserves are kept within a healthy range.

Why would the body build up resistance to the very hormone that makes life possible? Why would evolution tolerate the accumulation of genetic variants that cause insulin signaling to go so terribly wrong? **This is really not a case of bad genes, but rather of good genes forced to do bad things under abnormal circumstances.**

Insulin levels go up after a meal because the hormone tells the body systemically how to deal with all the new glucose, but the levels cannot stay up indefinitely. While there is a tap on insulin production at the source in the production, it is more efficient to control usage of the hormone at each tissue. This provides a type of buffering.

It might help to think of insulin resistance as caller ID. Telephones are great things when they help us remember to pick up some tomatoes on the way home or let us talk to loved-ones thousands of miles away. But when telemarketers worked out that they are also an effective device for hawking unwanted banking services just as we sit down to dinner, telephones became a threat. So we evolved caller ID so that we can locally screen the incoming messages, thereby making a relatively simple communication device somewhat complicated, but much better regulated and controlled.

One of the wondrous features of the insulin buffering mechanism is that it is flexible enough to adjust its sensitivity as life unfolds. During puberty or pregnancy, for example, we have different physiological responses to eating and adjust glucose metabolism accordingly, in part by elevating insulin release. Similarly, what is ordinarily a good level of insulin to get the job done, may not be so good as we go through phases of weight gain and healthy exercise, or illness, or just getting older. So the body is constantly adjusting levels of insulin resistance in proportion to the production of insulin in the first place. As insulin production goes up, so too does resistance. This sort of feedback loop serves to keep the whole system in equilibrium.

In obese people with constantly high levels of something called non-esterified fatty acids and with large rolls of abdominal fat deposits, long-term increase in resistance occurs, and the body enters something of a cry-wolf situation. Used to receiving the insulin signal all the time, eventually the person's cells become insensitive to it. At that point, resistance eventually leads to impaired insulin release in the pancreas as the beta cells shut down. What normally functions as a negative feedback loop to keep glucose in the normal range becomes the source of disease after a lifetime of stress. The genes aren't bad; they're just trying to do the right thing, but are getting it wrong because they are out of equilibrium with their environment.

Among the 50 or so genes thought to contribute to diabetes in this way, three are well enough characterized that even the most hardened skeptics would have difficulty refuting their involvement.

Calpain 10 is the textbook example both for how to find a gene for a complex disease and for differences among populations in the effect it has. Initially identified in Finland and quickly confirmed in Mexican Americans, for whom it is responsible for as much as one-tenth of disease susceptibility, it is nevertheless not involved in diabetes in Japanese, Samoans, or Africans.

PPARG has such a long formal name that it even puts biochemists to sleep: *peroxisome proliferator-activated receptor-gamma*. It encodes a growth factor receptor involved in regulation of adipocyte (that is, fat cell) development, whereas *Calpain10* encodes a protein that digests other proteins and affects pancreatic function. *PPARG's* oddity is that the susceptibility allele, thought to be a change of the twelfth amino acid from one type to another, is the common type in humans. More than three-quarters of us have what simplistically might be called the bad gene, but it has such a small effect that it does not lead to diabetes for most of us. On the other hand, most diabetics have it, so in the end it explains quite a bit of the disease susceptibility.

A third gene only just emerged from a series of whole genome scans late in 2006. It is evidently such a bad gene that it gets a criminal name: *TCF7L2*. Call it the Lex Luthor of genes, because it is also a mastermind, encoding a transcription factor whose role it is to control other genes. Like the other two genes, variation in *TCF7L2* accounts for as much as 20 percent of the incidence of diabetes in Africans and Europeans. Even more telling, the old, risky, allele has almost disappeared from east Asia, where correspondingly diabetes is relatively rare.

debunking the thrifty genes hypothesis

We now need to address the issue of why risk alleles are found at a common frequency in the human genome. There are logically three possibilities. First, they are just drifting around, too inconsequential for natural selection to pay much attention to. Second, a balance between the advantages and the disadvantages they confer may actively keep them around. Third, some of the alleles may be truly beneficial, but they are too young and there has not yet been enough time for them to displace the old ones. Evidence is accumulating that all these mechanisms are involved.

Let's start our discussion, though, with an explanation that you might have read about in the Sunday papers, the so-called "thrifty genes hypothesis." To be thrifty is to be wise with your resources, to be economical and not wasteful, to plan for the future. It follows that thrifty genes would be ones that allow you to conserve your food reserves for a rainy day, or another cache of berries or kill of wild meat as the case may have been for an early human.

It is precisely this concept that the University of Michigan geneticist Jim Neel had in mind back in 1962 when he coined "thrifty genes" as an explanation for the recent upsurge in obesity. Individuals who are better able to convert a surplus of calories into fat reserves would be more likely to see through times of famine that were presumably common as the species was first evolving. Yet confronted with the supermarket diet rich in sugars and fats and all things in plenty, those genetically same individuals are now predisposed to develop unwanted fat reserves.

Professor Neel was one of the world's leading human geneticists in the latter half of the twentieth century, but he had a case of foot-in-mouth disease at the end of his seminal paper. In a section titled "Some Eugenic Considerations" he made the precious remark that:

> If…the mounting pressures of population numbers means an
> eventual decline in the standard of living with, in many parts
> of the world, a persistence or return to seasonal fluctuations
> in the availability of food, then efforts to preserve the diabetic
> genotype through this transient period of plenty are in the
> interests of mankind.

He concluded: "Here is a striking illustration of the need for caution in approaching what at first glance seem to be 'obvious' eugenic considerations!" We might conclude that here is a striking illustration of the need for prominent geneticists to shut the heck up when they start speculating about and proposing genetic remedies for the future of mankind.

He should have known better. Food scarcity is going to cause a lot more pressing problems for humankind than preserving genetic variants, as the situations in places such as the Sudan and Rwanda indicate. Just how he envisaged the West implementing this engineering of thrifty gene prevalence in developing countries is disturbing to contemplate. We should give him the benefit of the doubt, though: Perhaps he was writing in late May as gray winter in Ann Arbor was finishing its eighth straight month, a stress mere mortals cannot be reasonably expected to handle with grace.

Lamentable eugenics does not however invalidate the thrifty genes argument, which seems to be thriving. Humans have a magnetic attraction to simple ideas that seem to have great explanatory power without a whole lot of tangible evidence in their favor. All the bits and pieces to the argument seem to make sense. All three premises of the theory of natural selection are met: There is variation among individuals, this is somewhat heritable, and there is differential survival related to the trait. Put them together, and you arrive at the conclusion that genetic variants predisposing to obesity must have been positively selected in modern human history.

Alas, close reading uncovers three features that a lot of evolutionary medicine arguments have in common: hyperadaptationism, the hereditarian fallacy, and sloppy quantitative reasoning. *Hyperadaptationism* is the notion that if an organism exhibits a trait, the trait must have evolved for that purpose. We now know that often traits evolve as a side effect of something else. Rather than assuming adaptation, evolutionary geneticists now accept a high burden of proof in establishing it.

The *hereditarian fallacy* is the idea that if there is genetic variation for a trait, and the frequency of the trait varies among populations, then genetics must account for the differences between the populations. It is that old racist curveball that keeps coming up in discussions of American IQ, raised here as evidence that selection accounts for the high prevalence of obesity among Pacific Islanders. None other than Jared Diamond, author of two brilliant books about the rise and fall of civilizations

(*Guns, Germs and Steel* and *Collapse*), has even gone so far as to suggest that Polynesian founders were likely to have been especially enriched for thrifty genes; otherwise they would not have survived the voyages across the Pacific. But Captain Bligh and his *Bounty* men were lean exemplars of the British Navy and survived their 49-day postmutiny ordeal, and there are numerous modern examples of survivors of calamities twice this long, so such inference of strong selection must be taken with a grain of salt.

Quantitative reasoning needs to be carefully applied to any proposal for adaptation, and on close examination the thrifty genes hypothesis falls apart. We can calculate what sort of benefit a genetic difference must afford for it to arise in an individual and be found in one-fifth of people a few thousand years later. It is in the vicinity of five percent, meaning that people with the advantageous allele have a 1 in 20 better chance of having children than those without it.

However, a thrifty genes critic from Aberdeen, John Speakman, has demonstrated that famine survival is highly unlikely to be anywhere near this strong. Basically, famines only rarely kill more than a few percent of the population in any generation, when they do they preferentially affect the very young and very old and so have little effect on genetic transmission. In any case, usually people die of infectious disease, not starvation. So we should be skeptical about claims that diabetes is the result of genes for weight gain being advantageous to nomads and pastoralists but bad for modern urbanites.

One gene actually does fit the thrifty criteria, but it is responsible for lactose tolerance, not weight gain. Prior to the domestication of goats and cattle, humans would not have drunk milk after weaning, or eaten cheese, yogurt, or any of those other great dairy products, as adults. The reason why many of us can digest lactose as adults is because we've converted a baby gene into an adult one. The enzyme lactase-phlorizin hydrolase (LPH or lactase for short) is used in the small intestine to allow babies to turn lactose into glucose and other sugars. On at least two different occasions in the past 10,000 years, once in Europe and once in East Africa, novel mutations have arisen that allow the gene to keep on working in adults. Population geneticists can estimate that possessing the mutation would have conferred something like a five percent fitness advantage to individuals in early pastoral societies that came quickly to depend on milk as a staple.

By contrast, **the most compelling argument that diabetes susceptibility didn't get into the gene pool a few tens of thousand years ago actually comes from the genes themselves.** In just about every case, it turns out that the ancestral allele, the one we share with chimpanzees and other primates, is the one that is more risky. In other words, the protective types are the ones that have been increasing in frequency for the last few millennia. Without these, obesity and diabetes would be even more prevalent than they are. Frankly, we have no idea yet what forces, if any, are favoring them, but we can thank our lucky stars that they are around.

disequilibrium and metabolic syndrome

A possible explanation for the increase in resistance factors is that there has been selection against the disease of diabetes. This seems unlikely since until recently it was not prevalent enough to affect reproduction with the necessary selection intensity. Perhaps there is selection involving a more subtle aspect of diet, and the impact of diabetes incidence is a side effect of this. Alternatively, the regulation of metabolism may be so complicated, involving hundreds of genes each with a variety of alleles, that it is inevitable that the network malfunctions in some people after a few key components of the system change.

One of the central ideas of this book is that humans are now outside their normal buffering zone. Instead of thinking of some optimal body mass or blood glucose level, we ought to be thinking of well-buffered systems of interactions that keep weight and energy within a critical range in the face of a constantly shifting environment. On this view it is absolutely critical that there be genetic variation to absorb the environmental insults that any organism faces: seasonal food sources, droughts, famines, illness, pregnancy, and growing old. They all put pressure on our metabolism, and it pays to have a flexible response.

For millions of years primates had a relatively constant range of pressures and evolved a system of hormones based around insulin, leptin, and some others that worked pretty efficiently. Then we decided as a species to start migrating around the world, to have odd monthly menstrual cycles, and to live longer than we were ever meant to. At various times we've switched from being herbivores to carnivores to omnivores, from hunter-gatherers to pastoralists, and most recently to cornivores. Our metabolic systems are stressed and confused.

Disequilibrium. Imbalance. Desynchronization. Instability. Mismatch. Call it what you will, but the fundamental problem is that our modern lifestyle is out of step with the genetic legacy of millions of years. Our omnivorous diet exposes us to a much wider range of toxins and pathogens than most species see, putting pressure on the exquisite network of cytokines and other signaling molecules that regulate fat and sugar metabolism.

Our wanderings have exposed us to such a wide range of climates and food sources that the metabolic system is forced to adapt locally. All these pressures and others are nudging the metabolic genetic network away from a balance forged over the course of mammalian evolution. **More than a third of all people living in developed countries are now at risk for a metabolic syndrome of ill health that includes diabetes and heart disease.** It will take tens, if not hundreds, of thousands of generations to find a new equilibrium to cope with the impact of a few thousand generations of profound perturbation.

Disequilibrium also exists between the energy dense Western diet and our genetic constitution, however, it has been shaped by human evolution. When you push any well-buffered system to the limits, it breaks down. We push the biochemistry and physiology of glucose homeostasis to the limits every day, and it is inevitable that some genetic combinations are less able to cope than others.

What to do about it? Western practice is to treat the symptoms. More drugs, please! Eventually, perhaps too society will call for eugenic approaches that will rid the gene pool of this scourge. Surely we will recognize instead that it is much easier and more human to change attitudes and shift lifestyles. Change what we eat and how we eat, and change our parenting practices that permit or even encourage young children to adopt the very habits that threaten their future happiness.

4

Unhealthy hygiene

athletic asthmatics Asthma is a modern disease that is the largest source of morbidity for children in the developed world, but happily can be overcome with appropriate attention.

inflammation and respiration Respiratory illnesses are an example of inflammatory diseases that result in part from inappropriate function of the immune system.

the hygiene hypothesis Part of the reason for the rising prevalence of asthma may be the great improvements in cleanliness and hygiene that characterize modern life.

asthma epidemiology However, hygiene is certainly not the whole story: Geography, socioeconomic status, and genetics all play their parts.

genetics of asthma Risk factors include immune regulators, genes involved in muscle function in the airways, and numerous others whose functions remain to be worked out.

inflamed bowels and crohn's disease Inflammatory bowel syndromes have similar genetic attributes and also seem to be exacerbated in the modern environment.

rheumatoid arthritis Arthritis is yet another inflammatory disease, of the elderly rather than the young.

imbalance of the immune system These diseases all reflect the imbalance between the recently evolved human genome and the constantly changing contemporary environment.

athletic asthmatics

A small boy passes his night crunched up in fits on his bed, chest to knees and forehead to pillow, just trying to suck enough air into reluctant lungs to feel comfortable enough to sleep. This incidence of mild asthma was brought on by who knows what, perhaps the tobacco smoke seeping out of his father's study, maybe the cat fur balled up under the old couch in the living room, or even something as innocuous as a mold hiding in the black bean sauce from last night's Chinese take-out.

Not that asthma is a new ailment: A long list of prominent figures in the Arts and Sciences testifies that it is an obstacle readily overcome by those of firm purpose. Politicians and revolutionaries of various ilk have suffered. On the one hand, the Central American activist Ché Guevara and Peter the Great of Russia; on the other an extraordinary group of five of the last 16 United States Presidents, including Teddy Roosevelt and John F. Kennedy.

Dozens of Olympic champions and professional athletes have brushed aside severe asthma en route to their ascendancy. Amy Van Dyken, winner of four gold medals in the pool at the 1996 Atlanta Summer Olympics, may be the most impressive example. Apparently her childhood affliction, induced variously by infection, allergies, and exercise, was so severe that on bad days she could barely laugh, let alone climb a flight of stairs. At the age of 11, she could not yet swim the length of a competition pool, but a dozen years later she was the fastest butterfly and medley swimmer in the world. It is hard to imagine two events more demanding on the lungs than these—except perhaps for the 400-meter freestyle. The Thorpedo, Ian Thorpe, one of the greatest middle-distance swimmers of all time, also took up the sport in an effort to increase his lung capacity to overcome childhood respiratory problems. He even had to overcome an allergy to the chlorine in pools, all without the help of steroidal antiinflammatories, which, of course, are on the list of banned substances for elite athletes.

On the track, Jackie Joyner-Kersee presents a different case study. In 1993, at the height of her career as one of the most dominating heptathletes of all time, she had an attack so acute that it almost took her life. She says it was like someone shoving a pillow over your face, completely blocking off the supply of oxygen. Jackie was not actually diagnosed with the disease until she was a freshman in college, having denied that there was a problem and hidden it from her coaches until then. After several trips to emergency care, she finally accepted the need for medical treatment and started down a path not just of recovery but also of athletic improvement. That she mastered the grueling two-day challenge of hurdles, high jump, shot-put, sprint, long jump, javelin, and middle-distance running should serve as an inspiration to anyone who feels grounded by the disease.

Jerome "the Bus" Bettis, lovable burly running back for the Pittsburgh Steelers, has a similar story of a turning point in his approach to the disease following a near-fatal attack during a televised game in 1997. His problem was diagnosed in high school and dealt with well enough that he thought he had outgrown it, so he relaxed his attention to the medical program. He fell into the bad habit of taking his inhaler only when he thought it was necessary, but says that this complacency ultimately led to his brush with death. Since then he has faced the disease with the same tenacity and force with which he hit the holes opened up by his offensive line over a decade of bruising football.

Childhood asthma is much more common than adult-onset, and most kids grow out of it to some degree, but it is a chronic disease possessed by 17 million Americans. This number is projected to double in the next 20 years, as it did over the preceding two decades. Five thousand deaths, 500,000 hospitalizations, and almost two million emergency room visits annually tell the tale of a major public health problem. Most attacks are said to be avoidable if only people pay attention to warning signs and stick to their medications. These include corticosteroids to control chronic inflammation of the airways, and bronchodilators to be taken prior to exercise or exposure to other triggers. Van Dyken, Joyner-Kersee, and Bettis are now three of the most prominent public figures of the Asthma All-Stars program dedicated to spreading the word about how the disease can be controlled.

inflammation and respiration

Inflammation is the body's natural response to a crime against it. Whether the agent is an insect bite or the sting of poison ivy, the aggressive swelling around a nasty cut, or an allergic response to house dust mites, there are commonalities in the way the immune system responds. When a crime occurs, our first response is to cordon off the area. Then we send in various experts to deal with the problem: police, emergency medical technicians, forensic detectives, clean-up crews. The crime scene becomes a hive of activity, set apart from regular day-to-day life.

So too with inflammation. The word comes from the Latin to "set on fire." It refers to an acute reaction characterized by redness, swelling, heat, and pain. These are things that follow from the opening up of the blood vessels at the scene of the crime where red and white blood cells are given access to the injury. If the skin has been breached, they initiate a cascade that will close the wound with a blood clot. If the wound has an allergic trigger, it will cause a big red rash, which may become an itchy skin condition known as atopy, eczema, or atopic dermatitis if you want to be really technical. Milder irritation of the airways gives rise to asthma.

One of the fascinating things about atopy is that it has increased dramatically in prevalence over the past few decades, both in dogs and in human children. The first recorded cases of canine atopy appear in the scientific literature just 50 years ago, but it is now estimated that upwards of ten percent of all dogs have irritable skin conditions. Atopy is closely associated with asthma in humans, but asthma is almost unheard of in dogs, while quite common in horses. Genetics predisposes in strange ways.

Hay fever is perhaps the most common inflammatory disease, a classical case of overreaction to a perceived threat. You can think of hay fever as asthma of the nose. North Carolina is covered with beautiful shortleaf and loblolly pines that grace its golf courses and lately frame its subdivisions. Every spring they blanket everything—cars, roofs and unsuspecting eyelids—in a bright yellow pollen, that if millimeters thick on the ground is predictably obnoxious to sinuses.

Confronted with unwanted pollen, circulating molecules known ignominiously as IgE strike up the alarm by signaling mast cells to secrete histamines at the site of intrusion. Both IgE and histamines are generally good guys, the former doing its best to serve as advance warning scouts for infection, and the latter helping out with other vital functions such as sleeping and having orgasms. But in pollen season they can be a nasty combination for those with a particularly sensitive disposition.

Asthma itself is a chronic condition where a person is at constant risk of having an attack triggered by the slightest perturbation, whether contracting a virus, reacting to a change of temperature or pollutants in the air, or stressing the lungs while exercising. Attacks are generally indicated by tightening of the chest and shortness of breath. It is like try-ing to breathe with someone sitting on your chest. Then as the inflamma-tory process builds up, mucus lines your bronchi and characteristic wheezing appears, sometimes accompanied by a persistent cough. Untreated, attacks can so severely cut off the oxygen supply that a patient's extremities will turn blue and cold, before they pass out, and possibly die from asphyxiation.

One big difference between asthma and hay fever is that the response to allergens is mediated by the more sophisticated T-cells, rather than just by IgE. Whatever is in the air that irritates an asthmatic is actually digested by professional cells whose role it is to present a chopped up fragment of the invader to the T-cells that mount a full-fledged inflammatory response. This is similar to what happens in type 1 diabetes, except that it is a foreign substance rather than insulin that is being presented to the immune system. And the immune system attacks the infected cells in the upper lungs or nose (rather than the pancreatic cells that make the insulin). In general it is a good strategy, but some people are more prone to mounting an unnecessarily severe response. Understanding why is the key to appreciating why there is genetic varia-tion affecting asthma.

Short-term relief during an attack is generally provided by bron-chodilators delivered through an inhaler. These drugs stimulate the air-way muscles to relax. Albuterol is the best known. The same drug is also taken intravenously to prevent premature childbirth since it relaxes the uterine muscles just as effectively. Ten or so similar drugs are on the market, prescribed according to strength, length of activity, and whether they cause tremors as they act on other muscles in the patient.

Long-term relief comes in the form of corticosteroids, which act to dampen the likelihood of an inflammatory reaction in the first place. As anyone who follows sports knows, steroids are a mixed blessing. They may have immediate benefits, but taken over a long period of time they are a risky business (quite apart from the potentially career-ending con-sequences of being caught). They are known to upset glucose metabo-lism and increase weight gain, so predispose to diabetes. They also promote osteoporosis by weakening bone development, and through their impact on muscle mass can so change the heart that they can cause cardiovascular failure later in life. It just doesn't seem such a good idea to take something that grows facial hair on women and gives men biceps rounder than their thighs, but in a way, this is what chronic asthmatics are forced to do.

the hygiene hypothesis

Of course, it would be better if we could prevent kids from being exposed to the allergens and toxins that are setting off asthma in the first place. Imagine a world without cigarette smoke and diesel fumes for breakfast, and we're probably halfway there. Yet according to an increasing body of thought, that same imagining of a world less prone to inflammatory disease would also encourage us to expose our toddlers to all sorts of viruses and bacteria, and to deliberately infect sensitive adults with intestinal hookworms.

The essence of this idea is embodied in "the hygiene hypothesis." First proposed by a British physician, David Strachan, in 1989, the hygiene hypothesis suggests that a large part of the increase in allergic inflammatory diseases in the twentieth century is attributable to the unusually sanitary conditions in which we now live. The immune system requires a delicate balance of efforts to fight diverse viruses, bacteria, and parasites. This must be primed during childhood, but since infants are no longer exposed to a traditional barrage of pathogens, they don't get the balance right. Hypersensitivity to unusual allergens follows, like weeds after a cleansing rainfall.

The idea is simple and compelling. Trouble is, there is precious little evidence that it is correct, at least as a general explanation for the epidemic nature of asthma. Like the thrifty genes hypothesis in Chapter 3, "Not so Thrifty Diabetes Genes," a great concept seems to succumb to the belligerent absence of evidence. Except that in this case, I'm willing to wager that the idea is right, and we just haven't worked out how to prove it: There is equally little evidence that the hygiene hypothesis is wrong—particularly since it is proving to be a remarkably adaptable hypothesis. The initial observation led Dr. Strachan to propose that the incidence of hay fever and eczema decline as families get larger. Furthermore, younger kids are less likely to suffer from these diseases, suggesting that they have been protected by exposure to all the ugly microbes that their older siblings brought into the house. Fifteen years later, Strachan used expansive hospital records to test his prediction that the decline in family size in England correlates with the ongoing increase in asthma prevalence. The results were equivocal, so it doesn't look like this simple form of the hypothesis is all there is to it.

Others have tested related predictions, also finding general consistency with the idea that improved hygiene increases the likelihood of asthma. Kids who attend day care and are exposed to a scary menagerie of bugs seem to be less predisposed to allergic illnesses. So too with kids who grow up on farms, rolling in the hay and sleeping with who knows what strange microbial bedfellows. Of course, they also grow up less likely than city kids to appreciate the difference between a cappuccino and a macchiato, a chardonnay and a pinot grigio, or ecstasy and heroin. Life is full of trade-offs that affect our health.

I have not seen it commented on anywhere, but the parallels between urbanization and asthma prevalence, and affluence and polio incidence are striking. Poliomyelitis is a viral disease that terrorized the United States and Europe for a couple of decades in the middle of the last century but has now been eradicated from all but the Indian subcontinent and tropical West Africa.

Its most famous victim was Franklin Delano Roosevelt, and its most enduring consequence has been the birth of biomedical philanthropism. Together FDR and his law partner, Basil O'Connor, marshaled heartstring appeals to the general populace to raise enough money to treat polio victims across rural America. Their March of Dimes Research Foundation was almost singlehandedly responsible for supporting development of Jonas Salk's vaccine that conquered the disease in America. The Foundation continues to support research into birth defects—in fact, they gave me my first research grant a dozen years ago. Today's Howard Hughes, Wellcome Trust, and Gates Foundations, among others, trace a direct line of descent to the March of Dimes proposition that philanthropy has a crucial role to play in curing common diseases.

The odd thing about polio, though, is that it was well known to preferentially afflict affluent white folk. It struck without warning, sometimes restricting itself to one highly athletic and active adult, at other times afflicting a whole family of children at once. We know now that the vast majority of infections are transmitted from feces to mouth, namely because of bad hand-washing practices, which is as good a reason as any to heed the signs in restaurant restrooms.

You would think that hand washing and modern style latrines would have been more common among the affluent 50 years ago. Perhaps some other aspect of personal hygiene was at play here. We also know now that only a few percent of all polio infections are symptomatic at all. The vast majority of the tens of millions of people who had the virus in their bloodstream probably never knew about it and lived perfectly normal lives without paralysis or gastrointestinal problems. Is it possible that the less affluent rural black population was protected because they had a stronger and better balanced immune system as a result of their life circumstances?

asthma epidemiology

What are the environmental factors influencing asthma prevalence? The distribution of asthma is remarkably inverse to that of diabetes. In the United States, levels are twice as high in the Northeast, Midwest, and Northwest, as they are in Texas and Louisiana, and they are intermediate along the Atlantic coast. An anomaly is Puerto Rico, where 1 in 5 kids are asthmatic, even though Hispanics as an ethnic group tend to have a lower prevalence than Caucasian and African Americans.

This geographic trend is not, however, reflected in measures of education and affluence. In just about every state in the Union, high school graduates have lower rates of asthma than those who drop out, while college graduates have significantly lower rates again. Interestingly, almost everywhere, not finishing college is accompanied by a spike in asthma susceptibility. The exceptions? Quite strikingly, California, where increased education is correlated with more asthma, and to a lesser extent the arid southwest states where the socioeconomic differences are not so pronounced.

This suggests that there are multiple layers to the impact of the human lifestyle transition from agrarian to modern urban settings on asthma susceptibility. While there quite likely is an impact of improved hygiene resulting in failure to appropriately prime the complex immune system, this is to a great extent offset by exposure to irritants. Chief among these are particulates in the house, such as cockroach or house dust mite dander, and fumes in the air, such as cigarette smoke, the fumes emanating from our freeways, and the airborne legacy of the industrial way of life.

Globally a couple of other phenomena confirm that the hygiene hypothesis explains only a fraction of the prevalence. Six of the seven countries at greatest risk of both childhood and adult asthma are Anglo-Saxon derivatives: Great Britain, Australia, New Zealand, Canada, the United States, and Ireland. The seventh is Brazil, whose latitudinal neighbor Peru is right up there as well, despite the fact that Portugal and Spain are if anything on the low side among European countries. These aren't just tendencies either; we're talking about whopping great tenfold differences. So prevalent is asthma in Australia that the folk cure of draping a towel over your head and inhaling Vick's Vaporub in boiling water inspired a scene in the movie *Crocodile Dundee*. Coming across a bleary-eyed man snorting cocaine, Paul Hogan intervened by dumping some of the white powder into a bowl of hot water and holding the unsuspecting partygoer's head over the fumes. A quarter of all Anglo kids have asthma symptoms, five times the fraction elsewhere.

The good news is that the world's most populous countries have the lowest asthma rates. China, India, Russia, Indonesia, and Mexico are well off in terms of incidence, though sadly they lead the world in terms of fatality rates among asthmatics. In fact, the fatality rate map is pretty much an inversion of the incidence one, with a disturbing exception. Among the afflicted nations, the United States lags notably in its ability to prevent fatalities. It doesn't take an epidemiologist to recognize that this has a lot to do with access to preventative health care for the less well off.

genetics of asthma

On top of all this, asthma tends to run in families. Not as much as blonde hair and blue eyes, but certainly more than cancer or alcoholism. Inheritance is often a telltale sign of genetics, but the inheritance of asthma might have as much to do with families sharing musty homes as predisposing genes.

The only way to really see whether genes are involved is to find them and work out how they work. Scans for genes shared by affected family members have been performed often enough that half a dozen places in the genome keep coming up, and another dozen or so have appeared often enough to make us suspicious. The conclusion is that asthma, like diabetes, is a pretty good example of a disease influenced by a complex mixture of dozens of genetic variants. Unfortunately, we do not yet know the identities of the genes in those bits of chromosomes shared by asthmatic relatives. The genetic terrorists, as it were, have been tracked to a few neighborhoods, but door-to-door searches will be needed to find the actual culprits.

That it can be done is shown by the example of *ADAM33*. A team of investigators at the threateningly named Genome Therapeutics Corporation tracked down this gene in hundreds of British and North American families. Just what is different about the version that makes some people susceptible to asthma is not yet clear, nor is it obvious how many of us are at risk, because any number of the SNPs in the gene could be responsible. (Recall that SNPs are the subtle changes in DNA spelling at a single letter—single nucleotide polymorphisms—that are littered throughout our genomes.) Nor is it obvious that knowing which type you have will be of much use to anyone in any case, since the relative risk it confers is pretty small. Nor does it account for more than a minor fraction of all asthma.

What is interesting about *ADAM33* is that it points our attention to quite a different aspect of the biology of problematic breathing than inflammation. This is because the gene encodes a pair of molecular scissors. Technically it is a zinc-dependent matrix metalloproteinase expressed in the bronchial smooth muscle and lung epithelium. Roughly translated, this means that it helps ensure that a person's airways are flexible.

In asthmatics, chronic inflammation causes the airway walls to thicken, which as well as narrowing the passages through which the air gets to the lungs, also inhibits the ability of the muscles in the airways to expand and contract. Those muscles can in turn weaken, which is why exercise is important for asthmatics to increase their lung capacity. The inflammation also locally damages the lining of the airways, so the body must remodel the lungs. This is where matrix proteinases like *ADAM33* are so important; they make sure that the appropriate balance between flexibility and rigidity is achieved.

Airway remodeling is also crucial for premature babies, many of whom develop severe wheezes as a result of the artificial breathing support they must receive. The unusual pressures placed on the developing lungs induce hormone pathways that help the airways recover. But part of the stress response is also to produce the very cytokines that promote inflammation, and which have been independently associated with childhood asthma.

Nevertheless, immune regulation is thought to be by far the most important genetic component to asthma. Many hundreds of studies have asked the question whether a particular allele of one of the dozens of key signaling genes is somehow associated with asthma, eczema, or some biochemical measure of hyperresponsiveness to allergens. The results are a mixed bag. One interleukin in particular, IL-13, has come up as a susceptibility factor in several studies, as has its genomic neighbor IL-4. Both of these are cytokines, little signaling proteins that push the immune system toward a state that is generally affiliated with asthma. The receptors that bind to these signals have also been implicated, as have various proteins that carry the message inside the T-cells. A couple of these are called IRAK proteins, not because they could possibly have something to do with the terror or hypersensitivity of allergic diseases, but because they are Interleukin Receptor Associated Kinases.

A couple of other genes are intriguing as well. One is a bit of an oddball member of our old friend the Major Histocompatibility Complex. HLA-G is not one of the picture hangers, but instead is probably involved in mounting the picture on the hangers—that is, the foreign protein fragments on the classical MHC receptors. There are two common variants, but whether they lead to asthma seems to be a function of whether the child's mother was asthmatic. If she is, and the child has two Gs at a particular place close to the end of the gene, then he is more likely to be asthmatic as well; if she is not, the same version of the gene may, if anything, be protective.

Mothers' immune systems have a tendency to see the developing fetus as something foreign, since it has a different genetic constitution after all, including bits from the father. Maybe the HLA-G expression helps her to avoid mounting an immune response against her baby. Probably too the baby ingests amniotic fluid that contains various cytokines that maybe reset the developing fetal immune system. There's a lot to work out, but it just highlights once more how difficult a job the immune system faces.

One of the high-tech genome-wide association studies has just appeared. The investigators looked at more than 300,000 genetic variants in 1,000 asthmatics and 1,000 nonasthmatics, and once again found suggestive evidence for a dozen or so associations with the disease. Some of these fall in the same regions of the genome as the bits of chromosome we discussed a the beginning of this section. But much the biggest factor is a SNP in the gene *ORMDL3*. Now that they have looked at more than 5,000 patients, this finding has a less than one in a trillion billion probability of being wrong by chance. You'll have to read the second edition to find out what *ORMDL3* does, because unfortunately no one has much of a clue at this time!

inflamed bowels and crohn's disease

Imagine now that instead of attacking a person's airways, allergic inflammation targeted the other end of the tube that runs down the middle of our bodies, namely the bowels. After a while, thickening of the walls of the intestine and colon would begin to restrict the flow of digested food, leading perhaps to diarrhea and almost certainly to abdominal pain. These digestive organs are full of bacteria, so it is inevitable that some microbes would get into the wounds and provoke ulcers and sores that might manifest as rectal bleeding, with anemia developing as the inflammation became more severe. It is a gruesome thought.

It is also daily life for about 1 in every 200 Americans and Europeans. Inflammatory bowel syndrome comes in two varieties known as Ulcerative Colitis and Crohn's disease. The differences are too subtle to concern us here; suffice it to say that they are about equally prevalent, and their incidence is rising. Let's hope it stays below one percent, because although these diseases typically alternate episodes of good health and relapse, there is nothing worse than chronic pain.

For some reason, Crohn's disease has attracted a lot of attention from the genome-wide association folks, and several studies have now clearly laid out the major genetic components. There are nine of them spread around the genome, each responsible for a few percent of susceptibility. A couple are clearly involved in regulation of the balance of the various arms of the immune system, and another couple are implicated in the ability to digest pathogenic intracellular bacteria such as *Listeria*. The others all seem to have roles in bowel functions and inflammations, but again the details remain to be worked out.

Like asthma, inflammatory bowel diseases are mostly maladies of the modern hygienic West. They don't appear on the scene until the 1940s, and then only outside the Tropics and predominately in white people. Plenty of theories are making the rounds—basically you could hypothesize a correlation with just about anything that we did to our world after the Second World War, such as watching television or driving cars—but a couple have a little more substance to them. One is the "cold chain hypothesis" proposed by a French team to explain their observation that Crohn's disease first became prevalent in North America, then England, and then southern Europe, each a decade or so apart. They blame refrigerators and the industrialization of food storage and production for enabling the spread of pathogenic gut bacteria such as *Listeria* and *Yersinia*. These bugs are perfectly capable of growing at low temperatures. There's a smoking gun in the form of the frequent observation of *Yersinia* species in the inflamed lesions of patients, though it must be said that it is far from established that these guys are causing the disease in any, let alone most, cases. Maybe there are other similar cold-loving bacteria we don't even know about yet that also ride the chain of chillers into our intestines where they aggravate at-risk people.

Another idea doesn't have a catchy name, so we'll call it the "latrine hypothesis." About a quarter of the human population is thought to be infected with hookworms, small centimeter-long nematodes that attach themselves to an intestine and gently gorge on blood. They are extraordinarily fecund creatures, each day shedding thousands of eggs that end up in a person's feces, and subsequently spread around pastures or ditches where people squat in the absence of toilets. When someone treads on infested excrement with bare feet, the hookworms make their way through the skin and over a few days migrate into the lungs where the larvae are coughed up and swallowed, commencing the life cycle all over again. The latrine hypothesis is the proposal that by using latrines on a regular basis, we have broken the life cycle of hookworms, which are thought to be highly protective against inflammatory bowel disease.

Relative to malaria and many other tropical diseases, hookworm infection is benign and in many cases asymptomatic. Ironically, where they occur, the symptoms of hookworm disease are similar to those of the inflammatory bowel diseases they are said to prevent. Hookworms do cause enough anemia and intestinal discomfort to significantly affect quality of life, so the Gates Foundation is supporting vaccination efforts in some parts of the world, such as the Amazon basin. Sewage disposal is even more effective and is the reason why the parasite disappeared from the southern states in the middle of the twentieth century.

Asthmatics and Crohn's patients alike are starting to consider the possibility that infecting themselves with hookworms might cure their ailments. The Web site www.asthmahookworm.com tells one man's desperate story of how he went traipsing in human excrement in the Cameroon as a last resort, and how he maintains that his nematode load alleviated his asthma symptoms. I doubt it would work for everyone and don't advocate his strategy, nor am I sure whether food supplements based on hookworm extracts are likely to do the trick, but we can hope that stories like this help lead to a cure.

Hookworms may exert a protective effect in several ways. One is that they are known to secrete an anticoagulant, which could conceivably counteract inflammation. Another is that they produce their own cytokines that might locally tilt the balance of T-cell responses in the walls of the intestine. Possibly their mere presence in the gut forces the body to mount a soft immune response that also alters the ratio of immune cell types sufficiently to offset the hypersensitive inflammation. For this reason, the cytokine IL4 has emerged from both genetic and immunological studies as a key player in inflammatory bowel syndromes and is currently undergoing clinical trials as a potential cure.

rheumatoid arthritis

For tens of millions of Americans, arthritis is a daily burden of chronic pain and discomfort, and the most pressing example of immune disease. We do not know whether or how hygiene and environmental change impact the incidence of arthritis, but once more it provides a compelling example of a complex disease that results from imbalanced activity of genes just trying to do their job.

My mother loves to start each day with a walk up the small mountain behind her home on the outskirts of Canberra. Whenever I return there for a short stay, I make a point of joining her on this ritual as soon as possible. The dry blue skies, familiar whiff of eucalyptus oil, squawks of magpies and kookaburra laughs, often even a glimpse of a pod of eastern gray kangaroos grazing in the underbrush, instantly reconnects me with my country and my childhood. Inevitably, Mum will point to something in the branches, perhaps a sulfur-crested cockatoo or a brightly plumed rosella, and my eyes will search in vain, having followed her outstretched finger bent ninety degrees in the wrong direction. Arthritis has gnarled that finger into a curl.

Why should over half of all elderly people experience this condition because of their genes? I do not want to give the impression that we understand what causes arthritis, but slowly the picture is coming into focus. The high prevalence is a little easier to understand once we recognize that arthritis is actually a suite of diseases that have similar symptoms. It is said to be genetically heterogeneous: Different genes are likely to contribute to each type of arthritis. The rheumatoid form, general inflammation of the joints, is thought really to be matter of friendly fire. The body attacks its own joints, mistaking these as sites of microbial attack. Some of the genes implicated in Crohn's disease are also susceptibility factors for rheumatoid arthritis.

Osteoarthritis is the other common form. It arises from wear and tear on specific joints as the bones grind over one another. Here the severity of inflammation is more like collateral damage as the immune system deals with bacteria attracted to damaged joints no longer bathed in their protective cartilaginous fluid. My guess is that old baseball catchers have a high rate of arthritis in their knees.

The few percent of people unlucky enough to have severe rheumatoid arthritis are victims of a triple whammy. First, they are predisposed to producing something unfamiliar on their cells as they get older. Then they have been exposed to some sort of an environmental trigger that uncovers that predisposition and causes the something to be produced. Finally, they are susceptible to recognizing it and overreacting. Rheumatoid arthritis is due to an adverse interaction between certain gene flavors, a change in the world around us, and some other flavors of genes.

The molecular players are likely to be different in particular cases, but one example now reasonably well established is as follows. All proteins are assembled from one of 20 common building blocks known as amino acids, one of which is arginine. It turns out that in the majority of rheumatoid arthritic patients, some of their arginines are converted into a similar-looking amino acid called citrulline. These are not to be confused with the foul-smelling lemon essence citronella that burns in candles or incense sticks to keep the mosquitoes at bay on a summer evening. After a time, this conversion of arginines to citrullines makes the patient's proteins look foreign to their own immune systems. One of the environmental factors that cause this to happen is tobacco smoke. Not content with causing lung cancer, emphysema, and asthma, cigarettes now have been associated with arthritis.

Some people, it appears, have genes that force them to present their new arthritis-promoting antigens to the immune system. As we have seen previously, these perfectly good and even essential genes are being forced to do something harmful specifically to humans in their modern environment. Aside from a fictional camel named Joe, other animals are not exposed to tobacco smoke. You just don't see many chimpanzees sitting around the Serengeti dragging on a pack of Marlboros. Those proteins that are presenting the citrulline on the surface of your cells do that for a living, but now they have been duped into delivering the wrong message.

Once more it appears that susceptibility to arthritis seems to be advanced in people who have a particular set of alleles in their Human Leukocyte Antigen genes. Recall that these are the baseball mitt-shaped picture hangers that present bits and pieces of molecules on the outside of our cells where the immune system can see them. To be precise, people with MHC Class II *HLA-DRB1 SE* are more likely to present the unusual homemade antigens such as citrulline, and hence to experience rheumatoid arthritis. This is just one of the many variable alleles in the HLA complex.

Inevitably, the stored up variation in the MHC can be led to do bad things. Another example is provided by the multiple sclerosis that afflicted the character Jed Bartlet on the television show *The West Wing*. For millions of Americans he served as surrogate President of the United States in the early 2000s, and the prospect of losing him to MS was as devastating as the glimpses we saw of progressive loss of central brain function that results in pain, depression, loss of coordination, vision, and many other symptoms.

It is likely that Jed carried the *DRB1°15* allele as one of his HLA flavors. Had his son-in-law been a medical geneticist rather than a much-ridiculed *Drosophila* geneticist, perhaps we would have found out. This flavor happens to present fragments of the protein myelin particularly well. Myelin is just one of the hundreds of thousands of fragments that *DRB1°15* can present. Normally it does not get the chance to do so because the T-cells do not cross something called the blood brain barrier. But in MS patients, immune invasion into the brain is triggered, perhaps by a bout of flu or high levels of stress, and a process begins that is analogous to that which causes arthritis in the joints.

The difference is that myelin is the substance that insulates the axons of your neurons, allowing signals to be communicated at high speed. Attacking it would be like someone sabotaging all the telephone lines that connected your home to the outside world.

The other really terrible well-known autoimmune disease is lupus. If arthritis is akin to friendly fire, then lupus is outright bullying at the hands of those assigned to protect you. Debilitating waves of attacks on the skin, joints, heart, lungs, blood vessels, liver, kidneys, and even the nervous system can strike at any age, and women are particularly vulnerable. Lupus apparently gets its name from the appearance of wolflike red blotches on the face, but that is as obscure as the disease itself.

Again the MHC region is implicated, but so are a dozen other places in the genome whose roles are yet to be defined. So too are stress, sunlight, and infection, and certain drugs can bring on the symptoms as well. The only good thing that can be said about lupus is that it is relatively rare and not obviously on the rise like so many other diseases.

imbalance of the immune system

The wonder with all of these diseases is not so much that they happen, but that severe forms afflict only a few percent of us. As with cancer and diabetes, we're dealing with an extremely complex genetic system in a species that has done an awful lot of evolving recently, in an environment that is fundamentally foreign with respect to most of our history. The question, "Why are there genes for arthritis?" is as misplaced as asking, "Why are there soldiers who fire on their own?" They don't mean to; it just happens.

Let's consider the complexity first. All organisms face a bewildering array of microscopic enemies. Belligerent bacteria, vile viruses, and pesky parasites confront the immune system with different challenges that must be met in cavities as different as the throat, intestines, joints, and sexual organs. It meets these challenges with immediate mechanisms such as coughs and sneezes, and an innate system that does things such as drilling holes in any bacteria that turn up where they should not be. Then it kicks in the so-called adaptive immune response, which has many layers and modalities depending on the specific needs. There is so much to coordinate that it is almost unavoidable that things will go wrong occasionally.

Some may say that this argument is a little bit of a cop-out, because adaptive immunity has been around for more than a hundred million years. Evolution has had plenty of opportunity to solve the riddle of how to distinguish self from foreign, so maybe you would think it would have perfected it. But as far as we know other mammals also get arthritis, degenerative encephalitis, lupus, asthma, eczema, and inflammatory bowel disease—support groups for dog and cat owners can readily be found on the Internet. It is just a downright difficult problem.

Next is the reality that optimal immune function is a constantly moving target. It is a bit of a myth to think that there is a single optimum value for any human trait, be it height or blood glucose levels or level of extraversion. It is definitely a myth to think that there is one optimal way to defend ourselves. This is most clearly seen for the MHC, which is the most variable part of our genomes, deliberately evolved that way to ensure that we can effectively fight off millions of potential pathogens. But it is almost certainly also the case for all the cytokines and other regulators of the immune response. There's an inbuilt balance of stratagems set to engage now for influenza and then for the common cold, and this is reflected in genetic variability in our genomes.

Whenever you have variability, you must also have individuals at the extremes. Most of us sit comfortably in the middle, but the price for this is that some of our friends and family are left exposed. They get combinations of genes that are probably ideal under some conditions of infection but leave them vulnerable in others. Apparently, one of these vulnerabilities is being hypersensitive to allergens that never would do much harm if left alone but look like they will, so end up inducing diseases such as asthma.

This is all exacerbated by the fact that humans occupy such diverse environments, many of which are extremely unusual from an historical perspective. What was good for a caveman is not necessarily good for an Eskimo, and what a Texas cowboy's immune system sees may not overlap much with the infectious exposure of a Michigan car salesman. As humans spread across the globe they put enormous pressures on their immune system, setting up a genetic imbalance that will be with us for hundreds of generations to come.

On top of which, we keep pushing the insults. Cigarette smoke and household dust mites are probably the two most nasty irritants in our daily environment: Neither has been a pervasive threat for more than a handful of generations. Add in diesel fumes and all sorts of other air pollutants, not to mention the pollens and perfumes of the new world, and you have a toxic mix for our airways to deal with.

Deal with it they must, but they probably do so from a position of weakness induced by the very conditions of life that have isolated us from the worst that nature sends our way. Hygiene itself, in ways we're only just beginning to understand, may be making it difficult for children to develop the robust and balanced immune systems needed to prevent inflammatory diseases. Deprived of exposure to pathogens that prime the development of adaptive immunity, it is thought that the army of T-cells just sits around getting soft or getting trigger happy. Oversimplified as this notion is, there is surely something to it.

Finally there is the matter that we are a young species, not yet in balance with our new surroundings, nor with our new human persona. At some places in the genome this is easily seen; at others we can as yet just infer that it is the case.

Precisely 524 letters upstream of the stretch of DNA that leads to the start of the *IL4* cytokine gene is the first of a run of five Cs. The sequence is identical in all of the great apes—chimpanzees, bonobos, gorillas, orangutans, and baboons. In fact, it is in a region of the chromosome that is pretty much the same in cats, dogs, and mice as well. However, if you're from Cameroon or China, the chances are pretty good that both of your copies of the gene read T instead of C at this 524th position. For Ethiopians and New Guineans, either type is equally likely, while folks from Italy or India most likely have Cs.

There are a couple of remarkable things about this state of affairs. One is that it is really unusual for the sequences of our genes to vary so much in a seemingly random manner. There are plenty of places where Africans differ from Asians, or Europeans from Native Americans, but for the most part these differences follow a pattern that traces the history of human migrations over the past 40,000 years. The C/T difference near the *IL4* gene stands out because the differences in frequency are much more regional. It provides a classical example of recent selection shaping the human genome.

Another is that people with the T version are able to produce considerably more of the IL4 protein. This is because another protein called NFAT prefers to bind to the sequence with the T, allowing the message of the gene to be read more often than when the ancestral C occupies the spot. This in turn makes such individuals just a bit more likely to be asthmatic and a bit more likely to succumb to leprosy and tuberculosis. These aren't obviously good things, but they are balanced by a slight reduction in susceptibility to different classes of pathogens for which one arm of the immune system is required, such as nematode infections and retroviruses.

IL4 is a wonderful and visible example of how the genome is in the process of responding to the different demands placed on the immune system as humans have migrated across the planet. A genomic tug of war is going on between a more or less active trigger, and it is likely playing out at several of the other genes implicated in inflammatory disease.

In parallel, the genome is also trying to adjust to the consequences of some of the other profound changes in human physiology that marked the birth of the species. With each new model of Ford or Chevrolet that rolls off the assembly line, subtle changes in the chassis hide subtle engineering feats that make the engine run more smoothly or the air conditioning adjust more quickly. Place the engine of a 1990 model in a brand new Mustang, and you would be pretty disappointed in the performance. Compare the climate control in a Taurus with that in a Camry, and you may get a sense that it can take time for new technologies to catch up with the body itself.

Evolution doesn't work as quickly as engineering. It is basically conservative, tinkering here and there, gradually improving, but necessarily the different parts change at different rates. Our lungs are pretty much built on the same design as those of other primates, but with some structural differences necessitated by our upright posture and changes in the larynx to accommodate voice. These developmental changes will have required subtle shifts in the genes that model lung development, including as we have seen, proteins such as *ADAM33* that give flexibility to the muscles that let us breathe.

A couple of hundred thousand years really isn't a lot of time to bring these genetic changes to a new equilibrium. The genome is coping, but it can do better. Given time and relative calm, we would expect that new genetic variants should spread through the human gene pool that make the lungs more efficient in the relatively new human body. Instead, we've added pollution and allergens that provoke inflammation, asking yet more of the genes, and have pushed them to the limits of their comfort zone.

5

Genetic AIDS

AIDS and the world   One percent of all humans will soon be infected, while as many as a quarter of the people already are in some countries.

from HIV to AIDS   There is no longer any doubt the virus causes immune deficiency, though coinfection with other pathogens causes the symptoms.

why HIV is so nasty   It is an insurgent that attacks the very immune system that is supposed to destroy it and a competent shape-shifter adept at avoiding drugs.

how to resist a virus with your genes   The surprisingly large amount of variation among people influences whether they become infected and how much virus they may have.

HIV imbalance   AIDS is a young disease and there has been insufficient time to build a genetic defense.

AIDS and the world

If you travel to sub-Saharan Africa or much of Southeast Asia, you will quickly find that our Western medical preoccupation with genetic disease is displaced by the manifest misery of infectious disease. Malaria sends a person through waves of debilitating fever, tuberculosis literally takes their breath away, and sleeping sickness and nematode infestation drain the lifeblood. That's not even considering by far the most prevalent pathogens on the planet: viruses. Gastrointestinal rotaviruses pose the chronic threat of diarrhea and dehydration for babies and children, occasional outbreaks of Ebola remind us that nature will eat us alive given half a chance, and even the common influenza virus leaves its scar on the world each fall. Perhaps the most pernicious of all, though, is the most recent arrival: Human Immunodeficiency Virus, or HIV.

Within a decade, one percent of all humans will be HIV positive. That's 60 million infections if you're counting, pretty much all within a 30-year period. Homosexuality can't account for this; nor can intravenous drug use, not even close. These were just the sentinel behaviors that triggered the world's attention back in the early 1980s. Far more accountable, at least on the global stage, is unprotected heterosexual intercourse, particularly that associated with sexual violence and prostitution. Now that more than ten percent of young adults are infected in countries such as Botswana and South Africa, the new threat is mother to child transmission, either during childbirth or via breast-feeding. Talk of losing a generation in the most destitute developing countries is no more scare mongering than talk of global warming. The threat is as real as the millions of orphan children who will soon inherit the villages and shanty-towns of southern Africa.

Much of the tragedy of AIDS, like many other diseases, is that it is preventable, but for human frailty. Conservative Christians are 100 percent correct in their claim that abstinence is the most assured way to prevent transmission of the virus, but their notion that it is achieved simply by placing trust in Jesus is as porous as used condoms, and about as effective. Al Gore in his *The Assault on Reason* talks of a rock-paper-scissors circle of faith, fear, and reason, in which faith trumps fear, fear trumps reason, and reason trumps faith. Here is a case where good intentions essentially throw the cycle into reverse, but with fear replaced by sexual desire. Add drugs to the mix, and the situation is nearly hopeless; even in the United States, crystal meth is breaking down the bonds of self-control that brought the spread of AIDS to a standstill a decade ago.

Life expectancy in southern Africa has dropped by ten years due to AIDS, in some countries hovering in the mid-forties. Leaders in government and the arts are being lost just as nations take tentative steps toward democracy. A powerful statement of this can be seen in the form of a permanent exhibit of dozens of exquisite stone sculptures that line the walk to the first terminal at Atlanta-Hartsfield Airport: Please step off the moving walkway and amble by them next time you have the chance. Meanwhile, the fabric of society is stretched to the limits as grandmothers resume the parenting roles of caregiver and provider for maybe a quarter of all families.

There is some hopeful news in that the prevalence of HIV infection is dropping in central Africa. Cynics bemoan that this may merely reflect high mortality rates of those first infected before anyone knew anything about the spread and control of the virus. Yet there is a crucial difference between countries such as Uganda where rates are at least steady and those such as South Africa where they continue to rise. That is aware-ness—awareness built by very public educational ABC programs pro-moting Abstinence, Being faithful (to a single partner), and using Condoms. Official policies that deny there is a problem or if there is one that it is due to an infectious agent, that promote herbal remedies in place of antiretroviral drugs, and that silence Western-oriented physi-cians aren't nearly so effective.

Highly Active Anti-Retroviral Therapy, or HAART, at $15,000 a year is an expensive but ultimately affordable lifesaver for half a million Americans. It is a cocktail of three or four drugs aimed at different aspects of viral biology. Hit with one drug at a time, the virus can easily evolve resistance, but hit with a combination it has little chance. For most people it extends life by a decade or more, and if mercurial basket-ball player Magic Johnson is any indication, in some cases it may even work to cure the disease indefinitely. In the developing world, generic local manufacturers have stepped in to provide the drugs for just $500 a year, but where incomes peak at $2 a day, even this is unbearable.

Drug treatment is wonderful for those who can get it, but it is not the solution that prevention will ultimately have to provide. Informed public policy, education, and chemical interventions with viral transmission are needed as well. The latter is where genetics makes its contribution.

from HIV to AIDS

I would hazard a guess that most Americans have never met a person suf-fering from AIDS. Perhaps they have heard about an old college friend after the fact, or know of a distant relative, but the closest they have come to the disease is through Tom Hanks's character, Andrew Beckett, in the movie *Philadelphia*: Remember the stoic conversations with his mother about T-cell counts, the harrowing operatic aria, the intolerance and fear mirrored in reactions to the purplish blotches of Kaposi's sar-coma, and the dramatic final courtroom collapse? Yet we never really learn exactly how the disease takes his life.

Unlike most viruses, HIV does its damage indirectly. By disabling the cellular immune system, it allows all manner of opportunistic infections to attack various organ systems. It basically opens the door to the pathogens that cause pneumonia and tuberculosis, diarrhea and colitis, fevers and encephalitis. AIDS is how we would all be without the T-cells that we met in Chapter 4, "Unhealthy Hygiene."

In fact, there are a dozen or so similar, and thankfully extremely rare, genetic conditions just like this, known collectively as Severe Combined Immune Deficiency. The immune system never develops in infants who have SCID diseases. The A in AIDS signifies that the disease is acquired rather than inherited, though viral transmission might as well be genetic when it is from mother to child, as is the case in a large proportion of HIV infection worldwide.

The probability of becoming infected from a single encounter with an HIV positive individual, either through unprotected intercourse of any type or by sharing of a needle, is actually quite low. But the less than one percent chance per exposure quickly translates into near certainty of infection when the risk is taken 100 times. The virus establishes itself by hanging out in places where the immune system is likely to be needed, notably around cuts and abrasions, where it can latch onto a passing macrophage or T-cell, which is why it is not a good idea to play contact sports when blood is in the offing. I was once refereeing a game of rugby in the early 1990s between a team from northern California and one from a mountain state, and asked a visiting player to leave the field to get bandaged up when he sustained a gash. He complained that this was unnecessary since they do not have AIDS in his town, but hightailed it to the sideline pretty quickly when I pointed out that they sure as heck did in San Francisco.

In the second month after a person becomes infected, the virus spreads through the immune system, seeding the roots of demise of lymphoid organs spread around the body. During this time, the patient may feel acute episodes of nausea, headache, rashes and sores, malaise, and weight loss, but it can take three to six months before infection is even detectable using biochemical tests. By then, HIV has gone into hiding inside the so-called CD4+ T-cell population. There it bides its time for maybe a decade or so, gradually eroding the numbers of these cells until they drop below 200 per microliter of blood, down from about 1,200 in healthy people. Only then is AIDS diagnosed, as occasional respiratory and genital infections give way to the typical course of immune deficiency.

So unusual is this pattern of disease progression that a few prominent virologists denied that HIV could be the cause of AIDS when the
virus was first isolated and shown to have infected hundreds of patients.
Peter Duesberg ran afoul of the establishment for many years with his
claims that the gold standard for establishing that a pathogen causes a
disease, Koch's postulates, had not been fulfilled. We now know that HIV
really is detectable in more than 95 percent of AIDS patients; the virus
has been cultured and shown to infect and inactivate T-cells; and extraordinary cases, such as accidentally infected nurses with no other risk factors, establish beyond a shadow of a doubt that HIV is responsible. No
one denies that other factors that affect the competence of a person's
immune system also impact the progression of disease, but denying
HIV's role is like denying that computer viruses cause software to crash.
Unprotected downloads usher in cyber disease, but unless they bring in
a virus, they won't bring down your system.

Without antiviral therapy, progression to AIDS is inevitable. A typical
telltale sign is the experience of a month or more of chronic diarrhea
brought on by bacterial infection of the gut. Thereafter upper respiratory
infections assert themselves, the most common of which are PCP, short
for pneumocystic pneumonia, and tuberculosis. **Together these
account for the majority of AIDS mortality, although drug treatments are used to control these, at least in Western countries.**
Inflammation of the esophagus is often due to infection with fungi or
common sexually transmitted viruses such as herpes and cytomegalovirus.
Further down the intestinal tract, a lot of bacteria that commonly live
with humans without causing too much harm cause gastric diseases and
eventually interfere with digestion, leading to characteristic wasting. In
up to ten percent of patients, still other stray infections with parasites,
bacteria, and viruses can attack the brain, perhaps leading to chronic
headaches, fever, and fatigue, if not AIDS dementia.

Cancer is another common cause of death in long-term AIDS
patients. Viral infection only rarely causes cancer in the general population, but coinfection with three common cancer-promoting viruses poses
a constant threat to HIV positive people. Kaposi's sarcoma is caused by a
herpes virus that starts by infecting the skin and mouth but may later
populate the lungs and intestines. Epstein-Barr virus will activate a particular class of lymphomas, knocking out the B-cell arm of the immune

system just when HIV has debilitated the T-cell arm. And women, who are particularly susceptible to coinfection with all sorts of sexually transmitted diseases, are prone to cervical cancer due to human papillomavirus (HPV). In fact, HPV is the target of a vaccine known as Gardasil that has recently become widely used by teenage girls. HPV is well known for its role in genital warts, but some strains are much more capable of promoting cancer of the cervix, particularly in immunologically compromised patients.

In different parts of the world, different infections dominate the progression of disease. Without a functional immune system, there is little long-term hope, and more often than not death is relatively slow and painful. An ever-present danger is that the pathogens will evolve resistance to drugs that physicians throw at them. Just imagine how it is to live in a society where this disease destroys the lives of more people than suffer from obesity or have blonde hair in our own society.

why HIV is so nasty

Ask why HIV is so elusive, and three compelling possibilities present themselves. First, the virus attacks precisely that which would destroy it, namely the immune system. Second, it is an expert at both camouflage and shape shifting, thus adept at avoiding counterattacks. Third, it is new to humans, who have consequently not had time to devise—genetically speaking—an effective strategy for coexistence. We will return to this last point at the end of the chapter, and address the first two reasons here.

It does not take a graduate of Annapolis to realize that destroying the enemy's defense complex might be a smart military move. Luckily for us, only a handful of pathogens have adopted this approach, HIV being one of them. It is not, though, a guaranteed strategy: Al-Qaeda tried something similar by piloting an airliner into the Pentagon, with spectacularly counterproductive results for hundreds of thousands of Middle Easterners. More successful has been the derived tactic of insurgency—particularly insurgency guided by infiltration of subversives into the police and army units. The human immune system does not have a central command, but it does have hundreds of command posts called the lymph nodes, which are precisely where HIV does its most dangerous work.

To infiltrate the immune system, HIV sneaks into T-cells, macrophages, and other important blood cells through the back door. Actually, it uses the front door as well. The relevant T-cells normally recognize foreign particles by virtue of a molecule called CD4, but each virus particle has a little hook known as gp120 that latches onto CD4 and uses it to open the front door and get inside the cell. At the same time, it must also engage with another molecule normally involved in receiving signals from the cytokines that regulate immune responses. These backdoor coreceptors are usually either the CCR5 or CXR4 molecules. The virus uses them to rearrange their own membranes to be part of the target cell, allowing their own genome to empty into it.

The viral genome consists of RNA instead of DNA. Almost always, RNA is the intermediate between genes and proteins, but in this exceptional case, it is both the messenger and the message. HIV is a retrovirus and carries with it an enzyme that reverses the normal flow of information by turning the RNA genome into a DNA copy, which then promptly jumps into one of the host cell's chromosomes. There it either lies low, or becomes hyperactive, pretending to be a regular gene by having many copies of itself made. The choice is a function of whether the T-cell is already engaged in an active immune response. In other words, HIV is so canny that it uses the fact that the host is already in a battle to ensure that it propagates and establishes the infection.

After the acute phase, when enough cells have been invaded and the immune response dies down, it doesn't have to do much, being content to sit and wait for the infected person to give it an opportunity to spread to someone else. In fact, it is probably in the virus's own best interest not to cause too much damage. Really nasty viruses that do kill their hosts in a few days are nowhere near as capable of spreading throughout a population.

So why is it so difficult for the immune system to do what it normally does, namely recognize the virus and kill it before it invades? Similarly, why don't vaccines work just as they did for polio and smallpox, tricking the immune system into action before a person becomes exposed to the real pathogen?

There is no simple answer, but part of it is that HIV is apparently able to camouflage itself by constantly changing its appearance ever so slightly. For example, at different stages of infection it may switch between which of the two back doors it uses. It can do this, because retroviruses have an unusually high mutation rate. Every one of the 10,000 or so letters in its genome has a better than 1 in 100,000 probability of changing each time the RNA genome is replicated. That does not sound like much, but it means that every tenth new virus particle is different from its parents. So what, you might say, every one of us is different from our parents at dozens of places. True, but there are only one or two of us for every parent, while a billion virus particles may be produced every day in an acutely infected person.

Natural selection among these viruses must be intense. So intense, in fact, that HIV avoids not just the immune system, but also many of the best drugs that humans have designed to deal with it. Early attempts to defeat HIV faltered because the virus just evolved ways around the drugs. The only truly effective strategy, as stated earlier, has been to employ triple cocktails of drugs. Reasoning that the virus is unlikely to simultaneously evolve three separate solutions, each ingredient in the cocktail has been chosen to target a different aspect of the way HIV propagates.

The cocktails typically involve a combination of an inhibitor of the enzyme that converts RNA into DNA, a "protease inhibitor" that blunts the molecular scissors that HIV uses to assemble its gp120 envelope protein, and unconventional nucleotides that substitute into the viral genome as it replicates. For example, AZT, also known as Zidovudine or Retrovir, was the first antiviral approved by the Food and Drug Administration for the treatment of AIDS. It looks a lot like thymidine (the T in the genetic code) but does not work like it. Similarly, Abacavir, ABC, or Ziagen, is an analog of guanine (the G in the code) that makes for a dyslexic virus.

Taken regularly and according to prescription, which is easier said than done, these combination therapies ought to extend the life expectancy of an HIV-infected adult from less than ten years to more than 30. Failure is often attributable to lapses in treatment due to lack of access to finances, clinics, or social support, but also reflects genetic variation for drug responses. Thus, a considerable proportion of patients are hypersensitive to Abacavir due to the specific variation in their MHC complex (the same genes that are so important in type 1 diabetes, lupus, multiple sclerosis, and other autoimmune diseases). We seem likely to be headed toward an era of personalized medicine in the treatment of AIDS, as for many genetic diseases.

how to resist a virus with your genes

You might think that since it is an infectious disease, genetics would not play much of a role in susceptibility to AIDS. This intuition turns out to be wrong, as in fact a person's genome can affect the course of disease in many ways. These include establishing complete resistance to the virus, setting the amount of virus in the blood in the years following infection, and influencing the impact of HIV on the rate of decline of the T-cell population. Certainly too there is a lot of variation for how resistant a patient is to all the pathogens that actually cause tuberculosis, pneumonia, or cancer.

A great example of this is the Δ32 mutation of the *CCR5* gene. Several years after AIDS was first recognized, it became apparent that a small number of Caucasian men were apparently immune to the disease. They engaged in high-risk lifestyles and were unquestionably exposed on multiple occasions to the virus, but never seroconverted to HIV-positive status. Some molecular sleuthing soon revealed that the reason for this is that they are missing a sizeable chunk of the back door that HIV uses to get into T-cells, the CCR5 coreceptor. The frequency of this mutation is up to twenty percent in northern Europeans, though it is essentially absent from Africans.

Something called the molecular clock allows us to infer that this mutation is probably new in the human genome. Population geneticists have sophisticated mathematical tools that they can use to date the time of origin of mutations by counting up how many differences in the DNA of the mutated gene make it different from "normal" copies of the gene.

It is a little like guesstimating how long a boy has had a broken leg by counting up how many scribbles are on his cast. A day after the plaster has set, he might have a couple of Jenny-was-here type scrawls from his sister, but weeks later the thing will be covered with limericks and all sorts of symbols and messages. So too with genetic mutations. A few generations after one appears, the chromosome is essentially identical to the original one, but hundreds of generations later it will have picked up more mutations. There is a lot of randomness to this molecular clock of accumulating differences, so the best we can do is infer an interval of time in which an event happened.

In the case of the *CCR5* Δ32 mutation, we're reasonably confident that it was somewhere between 300 and 1,800 years ago, most likely some time in the Middle Ages. An origin 600 or so years ago is pretty intriguing given that this is when the Black Death decimated Europe. Several authors have, in fact, argued that the mutation offered protection against bubonic plague, and this is why it is highly prevalent north of the Alps. This would be a great example of how our recent history of disease may have shaped resistance to a new pathogen, preventing us from getting sick.

It is a nice idea, but the details are almost certainly wrong. To see why, imagine a medieval village of 10,000 people in which, for one reason or another, the *CCR5* Δ32 has already attained a frequency of one percent even before the plague arrives at the town gates. Even at this frequency, there would likely only be one person in the town with two copies of the mutation. If HIV resistance is anything to go by, then only this individual would be completely resistant to the plague, but, for the sake of argument, suppose that even those carriers with a single copy are afforded complete protection. Suppose further that it is a really catastrophic plague and that one quarter of the townspeople die the horrible death. Then among the 7,500 survivors would be every one of the 200 or so carriers, and the frequency of CCR5 Δ32 would have risen to 200 in 15,000, or just one and a third percent. It would take dozens of such calamities for natural selection on this scale to get the allele anywhere near ten percent. Bad as it was, nothing like that scale of devastation occurred: Only two extreme plague outbreaks centered three centuries apart in 1350 and 1665 are documented.

A more likely culprit is smallpox, or some other hemorrhagic fever virus, that may have claimed the lives of a few percent of the population a year for 700 or more years. Others have played with the models and come up with plausible scenarios. If transmission is more likely within families than among the general populace, because of shared exposure and care giving, then any protected individuals also protect their relatives to some extent. This notion of inclusive fitness allows for a wide range of conclusions depending on the assumptions that are made, so it is difficult to be sure. In any case, it does seem likely that some modern Whites are at reduced risk of succumbing to AIDS thanks to the unconscious sacrifice made by millions of people hundreds of years ago in the face of the new urban plagues and fevers of the time.

For the vast majority of patients who do not have the mutant receptor, there is some hope that drugs might be used to throw a blanket over

CCR5 instead. Pfizer is currently carrying out advanced clinical trials for its compound, Maraviroc, which it is hoped will prevent the spread of the virus. Unless of course HIV uses its shape-shifting capacity to evolve to enter through the CXR4 back door instead.

What about those people who do become infected? Is there any-thing about their genetic constitution that might make them more or less susceptible to AIDS progression? Evidence is accumulating that there is, some of it coming from another one of the whole genome association studies that we met in the previous few chapters. It seems that in this case the geneticists need not have cast their net so wide, because our old friend the MHC is cropping up yet again—but in novel and unexpected ways.

During the quiet phase after a person converts to being HIV positive but before she starts showing the symptoms of AIDS, the virus is detected at a level in the blood that is characteristic of the patient. This level, called the *viral set point*, ranges over five orders of magnitude, meaning that some patients have 10,000 times more circulating HIV than others do. It stays that way for several years. Oddly, the amount of virus does not necessarily determine how quickly the disease will progress. It may, however, impact the likelihood that a person will trans-mit the virus to others.

Two major places in the MHC seem to provide some protection. Unimaginatively, as usual for human geneticists, one is called *HCP5*. It encodes a rogue copy of the enzyme that turns RNA into DNA in other retroviruses. There are signs that the whole gene is a new arrival in the genome, since it is found only in some primates. Worse, the protective allele is once again almost absent from Africans, and Asians as well. Only five percent of Caucasians have it, but for those who do, like *CCR5 Δ32*, it is something of a genetic condom.

Not far away is a much more common variant found in about half of all people that also affects the amount of virus in the blood. This allele appears to affect the utilization of a portion of the MHC that was not previously thought to play a role in recognizing any viruses, let alone HIV. Together with *HCP5*, it explains as much as 15 percent of the varia-tion among patients in how much virus they have in their blood. Presum-ably, if all of us had the alleles that are currently low frequency, then there would be a lot less AIDS.

Quite a way further down the chromosome is yet another variable site, this time in the *ZNRD1* gene that actually lengthens the time interval between infection and onset of disease. *ZNRD1* is what we like to call a housekeeping gene: It has a crucial yet somewhat dull job in every single cell. That job is to make components of protein factories. It is bizarre that this variant influences AIDS susceptibility, analogous to changing a minor component of all the local power relay stations around the country and then observing that we're now protected against a particular brand of terrorists. Perhaps I should wait until the result has been confirmed and studied more before trying to explain it any further.

The point is, though, that if genetic factors influence how much virus a person has in his blood, or how long a person can survive with the virus, there ought to be ways for scientists to work out how to manipulate the genes. Doctors might also be able to use the patient's genotype to guide when they ought to start administering the expensive drugs while minimizing the time that the virus has to evolve resistance. Without a doubt, education and sound public policy are the keys to dealing with HIV globally, but it is likely that new generations of drugs will play their role as well.

HIV imbalance

The youth of HIV as a human virus may be the greatest source of its threat to our species. Astonishingly, simian forms of the virus, SIV, are widespread in chimpanzees and some other apes, where they cause minimal morbidity. Similarly, the closely related feline immunodeficiency virus, FIV, is common and relatively benign in wild cats such as cheetahs. For some reason, canines are free of such viruses (one is tempted to say, on no grounds other than the pure prejudice of a dog-lover, because of their intelligence).

Tellingly, SIV and FIV both become deadly agents of immune disease the minute they cross into a nonexposed sibling species. Domestic cats that become infected soon succumb to AIDS-like symptoms, indicating that they do not have the antiviral defense mechanisms that have evolved over millions of years in their wild relatives. Macaques are the preferred animal model for AIDS studies because, like humans, they have not carried SIV during their sojourn as a species, and it causes them to become sick. Wait another few thousand generations and it seems likely that we too will evolve responses that allow us to live with the virus as a normal part of our biology, much as most humans coexist with CMV and probably many other viruses.

Current thinking is that HIV jumped from primates to humans on two separate occasions early in the last century. There are actually two different lineages of the virus. HIV-1 is the more pernicious form and is the common one outside Africa. It was probably contracted through an open cut in a person who was butchering an SIV-infected chimpanzee, and quickly evolved into the human form. HIV-2 may have arisen more recently from a sooty mangabey, since its sequence looks most like the form of SIV found in that species of monkey.

For a while, a conspiracy theory was doing the rounds, asserting that HIV crossed into humans during the preparation of polio vaccines from monkey kidney cells in the Belgian Congo in the early 1950s. The first documented case of a patient with the virus comes from around that time, in Africa, but this is likely just a coincidence. For one thing, the molecular clock method tells us that the virus was almost certainly in humans several decades before then; for another, an old vial of the original vaccine has no trace of it, and it turns out that only macaque kidney cells were used in making the vaccine, but macaques don't naturally have SIV.

The bottom line, in any case, is that AIDS is a very young human disease whose primary cause is a virus that is only now in its third or fourth generation of human exposure. Yet there is a wide range of genetic variation for how people respond, some of it reflecting our general history of disease, some of it possibly of little or no functional importance until now. The situation is little different from the other major common diseases that this book is concerned with, namely cancer, diabetes, asthma, depression, and dementia, but for the fact that the environmental exposure is to a virus rather than carcinogens, carbohydrates, pollutants, psychological stress, and unusually old age.

At first blush, it may seem strange there is any variation in our genomes for how we would respond to a new pathogen. One view of evolution supposes that species are so exquisitely finely honed, that they do not carry around any variation not needed for some purpose. Defending oneself against a virus that did not exist in all but a handful of people until 20 years ago would not seem to be much of a purpose. Proponents of this view would prefer that the variation had arisen after the threat appeared, as an adaptive response, waiting to spread through the gene pool. But there has patently not been sufficient time.

A different view has it that in fact the situation is the opposite, namely that the genome is full of variation of so little consequence that natural selection cannot be bothered with it. This is particularly apt for a species such as ours that has just come through a rapid evolutionary spurt and is equilibrating to all manner of new situations. On this view, it is to be expected that some of this variation all of a sudden assumes functional significance in the new environment. *HCP5* could be an example of this effect, a mutation that has had little consequence since it popped into the chromosome, but which now confers protection against HIV.

An even stronger version of this theory posits that, in fact, the genome has evolved to suppress the effects of some genetic variation. If you own a pool you might consider the pH tests you're supposed to run every few days, where the test solution is green for the first dozen drops you add from tube M007, and then suddenly turns red on the thirteenth drop. What is happening is that the extra little bit of perturbation pushes the chemical system past a threshold, and it switches phase.

Genetic systems are similar. They are so exposed to differences in the environment that they have evolved buffering mechanisms that minimize the effects of hundreds of subtle mutations that are in the gene pool. These subtle mutations drift around not bothering anyone until some major change in the environment pushes them outside the buffering zone, and now they influence the course of disease: Sometimes they may promote it; sometimes they protect us from it. In this respect, HIV is just a microcosm of the same types of processes that affect many other complex human diseases.

6

Generating depression

creative depression It is amazing how many world leaders and entertainers have suffered from the disease.

an epidemic of mood swings Depression is on the rise, afflicting well more than 1 in 10 people at some point in their life.

bipolar and monopolar disorders Each, with differing degrees of severity, are much more than prolonged sadness.

the pharmacology of despair Serotonin and cortisol are a neurotransmitter and a hormone that jointly regulate mood.

misbehaving serotonin Genetic variation in the serotonin transporter is associated with some measures of depression and suicide, particularly if life is stressful.

faint genetic signals It seems likely that hundreds if not thousands of genetic variants contribute to mood disorders.

schizophrenia and other mental disturbances Many other types of mental problems are equally as genetically complicated.

the genetic tightrope of the mind Cultural and genetic change have affected the brain perhaps more than any other organ.

a kindling theory in the modern world Depression gets worse with time as mood becomes less stable and more sensitive.

creative depression

When the "Piano Man" sings about men at the bar sharing a drink they call loneliness being better than drinking alone, the melancholy is somehow erased by something uplifting in the melody. Few people are aware that the lyrics of another of Billy Joel's anguished songs, "Tomorrow Is Today," were penned as a suicide note. For most of us, complete loss of the will to live is incomprehensible, yet more and more people experience the profound and debilitating sadness of clinical depression. What's going on and what might our genes have to do with it?

It is astonishing to consider how many creative artists have fought or fight the demons of sadness. When *Saturday Night Live*'s Chris Farley committed suicide, it came as a complete shock to the American public, but he is just one of a half dozen well-known comedians who have at least attempted to take their own lives. Contemporary funnymen Drew Carey and Jim Carrey express their infectious senses of humor in different ways, but for each, the public persona belies private swings of darkness. The bullet in Kurt Cobain's temple left little doubt that he was unable to find his personal Nirvana here on earth. Even a whole genre is named after the blues. And can it really be true that Harrison Ford, the actor behind the swashbuckling heroes Han Solo and Indiana Jones, got his start in acting in part because his inability to overcome depression-induced sleeplessness had him thrown out of high school?

Nor are intellectual and world leaders immune to the debilitating shadow of sorrow. Two of the towering figures of the nineteenth and twentieth centuries, Abraham Lincoln and Winston Churchill, both suffered from melancholy and depression. Some have argued that Churchill drew inspiration to overcome adversity and the forces of evil from his battles with what he called his "black dog." Few can argue that the illness can be a well of tremendous creativity: Witness the talents of such profound writers as Tolstoy, Goethe, and Kafka; consider the lives of Tennessee Williams and Ernest Hemingway.

This male-biased menagerie misses a crucial point, which is that women are in fact far more prone than men to mood disorders. If you read accounts on the Web, you quickly notice that depression in women is often attributed to situational cues. Lady Diana, Princess of Wales, battled bulimia in the glare of the enormous pressure of popularity while trapped in a loveless marriage. Brooke Shields and Marie Osmond are the poster women for postpartum depression. Amy Tan and Anne Rice are said to project the tragedy in their lives.

The reality though is that the disability has organic roots and casts a shadow that must be managed and confronted on a lifelong basis for tens of millions of typical adults, the majority of them women. As Sheryl Crow puts it in "Soak Up the Sun," she's "gonna tell everyone to lighten up (because she's) got no one to blame for every time I'm feelin' lame." There is no point in blaming the genes either, but as we shall see, these are part of the equation, once more because we're still finding our way as an evolving species.

an epidemic of mood swings

At any given time, somewhere in the vicinity of four percent of women between the ages of 25 and 45, and two percent of men, are clinically depressed. These are conservative estimates. They mean that without a doubt someone in your neighborhood or workplace is struggling mightily to face each day with characteristic cheer. Multiply them by five to obtain the number of people experiencing the blues, and double again to arrive at a lifetime estimate of the fraction of the population that experiences at least one somewhat debilitating episode of depression.

As far as we know, this disease recognizes no ethnic or socio-economic boundaries, though of course it is almost impossible to know. Many countries refuse to recognize the existence of mental illness, let alone assemble accessible mental health records. One thing we do know is that in the developed world, it is the leading source of lost life potential. It is expected to become the second leading source of disability worldwide, behind heart disease, in the next decade.

For this reason, depression is sometimes said to be epidemic, implying that it is more prevalent now than ever before, and maybe that it is spreading in an infectious manner. If there is an infectious agent, it is almost certainly cultural, not microbial or viral. Whether depression is truly on the increase is difficult to know. Some authors maintain that the prevalence is underestimated by as much as fifty percent, due to the stigma attached to mental health and the ends that people will go to in order to hide their socially unacceptable anguish. Others maintain that it is overdiagnosed to at least the same degree, because people who are just dealing normally with growing up or with trials in their lives are being incorrectly tagged as depressive.

Those who have never battled the disease often suppose that it is just an extreme form of sadness, but anyone who has so much as brushed against a prolonged episode of depression is more likely to recognize the deeply organic nature of the disease. Every person's experience is unique, but in the midst of the inability to sleep or to care or to just perform daily routines is the awareness that it is all so very wrong. You know that your behavior is self-destructive, that it is hurting friends and family, and that there is no rationally defensible reason for you to be so self-obsessed and sad. Yet you are absolutely powerless to purge the negative feelings. You can almost feel the bad chemicals swimming around your mind, imposing their will on your mood as surely as alcohol swamps your conscious well-being in a very different way.

For a chilling description of what a mental breakdown due to depression is really like, try Andrew Solomon's *The Noonday Demon*. Subtitled *An Atlas of Depression*, this is a superb account of the disease from the perspective of dozens of different patients, full of sanity and realism. It opens with a telling description of the author's own battle with the demons, from childhood anxiety attacks to the incapacitating depths of despair in the months after publishing his first novel. What a frail thing is the human mind, what a wonderful thing its capacity for compassion and understanding.

bipolar and monopolar disorders

Psychologists commonly recognize four subclasses of the illness. A basic distinction is made between unipolar and bipolar affective disorders, and within each of these there are at least two categories of severity. Since polarity by definition refers to orientation with respect to two opposites, the term unipolar makes about as much sense as a magnet without North or South, or the sound of one hand clapping. It is used to distinguish patients who experience only the lows of depression from those who alternate low with high episodes, the term bipolar supposedly carrying fewer stigmas than manic depression.

The least disruptive form of depression is a pattern of constant lack of interest in activities and inability to enjoy life that lasts for at least two years. The technical term is *dysthymia*, which is surely preferable to minor depression, because there is very little minor about it. Dysthymic individuals have problems sleeping, have persistently low energy levels, often either lose their appetite or have troubles with overeating, and may constantly fight feelings of helplessness and of poor self-image. The fatigue keeps them in bed for hours and hours into the day, yet it can also keep them awake long into the night. Doing things that come ever so naturally, such as making a cup of coffee, getting dressed, making a phone call, going for a walk, or even tucking the kids into bed become monstrous chores.

The way to navigate each day is to make each minute or hour a succession of mileage posts that are passed successfully, finding pleasure in the mundane and routine. At times the pressure may become so great that a person is effectively paralyzed, if not physically for minutes or hours, then socially for days or weeks at a time. It is a heavy burden to place on a career in this fast-paced world. It is a ghastly strain to impose on a marriage or relationship that started with such joy and promise. It is a debilitating tension on the psyche that sits constantly poised on the edge of complete breakdown. It is no way to lead a happy life, but it is the reality for 1 in 20 people today.

Major depressive disorder is worse. It is characterized by recurring episodes of pronounced inability to function in anything approaching what might be called a normal manner. It is worse than melancholy, deeper than grief, more than an exacerbation of the symptoms just described. Atypical episodes, which are actually more common, may include overeating and oversleeping, but can give way to leaden paralysis and heightened sensitivity. Normally calm people find themselves reacting with irrational irritability to slights such as being bumped into in the street or jumped in front of in a queue. These episodes of inability to react with positive feelings to anything last from two weeks to a month or two. Now days, chemical intervention is almost always prescribed.

Typical melancholic major depression involves loss of pleasure and a state of depression unlike anything most of us ever experience, a tendency toward anorexia, and such a slowing of thought processes that people are simply incapable of much more than lying around. Solomon describes being so low that he could not even contemplate suicide, the danger period for that and self-mutilation being during the recovery phase. In severe states, patients require nursing help with eating and essential bodily function.

Yet it can get worse, with the descent into psychotic depression. Symptoms include hallucination, aggression, feelings of intense hopelessness and frustration, loss of the sense of self, and possibly delusional or extreme paranoid behavior. I cannot imagine what it is like to live with the constant awareness that such a state may be just around the corner awaiting some trigger, tragic or prosaic, but always unpredictable, knowing too that the drugs that are effective in stabilizing the condition may not be able to hold back the tide of a psychotic episode.

Speaking of triggers, two of the more prevalent forms of disease are postnatal depression and seasonal affective disorder (SAD). The baby blues, consisting of sleeplessness, irritability, headaches, and impaired concentration, apparently affect more than three-quarters of new parents and last a day or two (actually, for different reasons, a decade or two). But some combination of an anxious and naturally melancholy personality, the stress of parenthood, perhaps exacerbated by an unsupportive marriage, stressful job, substance abuse, hormonal swings, and flat out chance, leads to at least five percent of mothers (and some fathers) entering an episode of clinical depression. Untreated, postnatal psychosis can result in infanticide, a generally unthinkable crime that challenges our sense of personal responsibility, and which society is patently unsure how to approach.

SAD by contrast is perhaps the best-named disease on the planet, although if the Icelandic term *skammdegisthunglyndi* rolled off the tongue a little better it could catch on as well. It means short-day-heavy-mood, and communities with a high prevalence know exactly what to do about it: bathe in the light of a sunlamp for half an hour every morning. My Seattle colleagues swear by it. Given a choice it seems simpler to live by a sunny beach. Citizens of the cloudy people's republic of Ann Arbor regularly prefer to imbibe the target of sunlight, melatonin, and fittingly wander around battling jet lag symptoms half the year.

Bipolar disease is also conveniently categorized in two types, I and II, which are differentiated by the severity of the manic phase. Mania refers to a period of elevated mood lasting for at least two weeks. The full-blown type I form is often associated with a racing mind as if the brain is working too fast. Sleep is disrupted, the person may be irritable or may be unable to pay attention, and delusions of grandeur are frequent, giving way to psychosis, namely a loss of contact with reality. Less than one percent of people suffer in this way, but an equivalent number if not more may experience the lesser form, hypomania.

Hypomania on the face of it does not sound so bad. Many of us would gladly experience the creative surges that accompany the state, allowing people to write poetry, compose music, have unusually original ideas, and overcome social inhibitions for a time. It sounds a lot like staying at a Holiday Inn Express overnight. I wonder whether that experience also blunts a person's emotions and makes a person laugh uncontrollably.

Typically, manic and depressive phases last a month or two and alternate every other year, but there is enormous variability. Some cycle much more rapidly, several times a year, and some experience both conditions simultaneously, which you can imagine makes for a volatile state of mind. Some have much stronger manic phases, while some have more prevalent depression. Particularly in children, diagnosis is complicated by coincidence with other psychological conditions including attention deficit hyperactivity disorder and schizophrenia.

One final feature that all modes of depression seem to have in common is that left untreated, the episodes tend to get more intense and more frequent as life proceeds. A first episode of major depression almost always indicates a lifetime of battles, with control, rather than cure, the therapeutic objective. A depressive's mind wanders along the edge of the cliff, prone to missteps that leave it clinging to sanity until rescued by a helping hand that can never quite drag it to the plane of normality.

the pharmacology of despair

Serotonin is e-mail for the brain. This one little chemical seems to be the key to the lousy mental state of a billion people whose neurons just can't communicate efficiently. One way or another it is the target of just about every antidepressant drug on the market, and quite likely of psychotherapy as well. Google "serotonin" with "depression," and you will find hundreds of thousands of hits. Search with the same two words on the academic search engine PubMed, and you will be led to more than 12,000 scholarly articles. This is a pretty good indication that the two are linked, but also that we really don't understand how.

It is all so very complicated. Serotonin does many different things with many different partners, and many different things do some of the things that serotonin does. Among those 12,000 articles you will find references to inducement of vomiting, the age at which we first have sex, sudden infant death syndrome, and aggressive lobsters. They will tell you that humans have seven different types of receptors encoded by 14 genes, and impart more exquisite detail concerning the biochemistry and pharmacology than even an expert can possibly assimilate. Look more deeply and you will discover that at some level most mood-altering drugs converge on serotonin: ecstasy and crystal meth, LSD and all manner of psychedelic snuffs, Ritalin and Fen-Phen, pituri and betel nuts, and indirectly even caffeine and nicotine.

So while low levels of serotonin activity are certainly causal in depression, simply bathing the brain in the chemical is not necessarily a particularly good idea. Actually, it is not that simple, either, since ingested serotonin cannot find its way into the brain. You can try eating foods rich in its precursor, tryptophan, which is said to help, but in my experience those who have diets rich in cottage cheese, wheat germ, and soy aren't necessarily happier than the rest of us. (Actually, poultry, fish, eggs, avocados, and other beans are also good sources.) The preferred method of serotonin enhancement is via a host of drugs.

To see how these work, let's return to the e-mail for the brain analogy. We're not talking about the content of the e-mail here, just the fact that it is a major means of communication. It facilitates talk between neurons, while the content of the messages is a function of which neurons are being connected when and how. Serotonin is not an information-carrying molecule, it is just a chemical that connects one neuron to another, encouraging the electrical signals to continue along their way. Many other chemicals have a similar role, and we can think of them as regular mail, telephones, and text messaging, but why interference of serotonin in particular leads to depression is simply not clear.

In any case, if you are using dial-up or have an unreliable Internet service provider, then your e-mail is slow, and these days most of us can't function properly without it. This is basically like not making enough serotonin, so drugs such as SAMe and 5HTP have been introduced because they fiddle with the enzymes that convert tryptophan into serotonin. For some reason, they are not yet particularly popular, even though clinically they can be effective and seem to have limited side effects.

Then, perhaps you are still using one of the old e-mail servers such as Pine or Eudora, in which case you can function but probably are not utilizing e-mail as effectively as you might. This would be the situation if one or more of the receptors for serotonin were not functioning properly, but unfortunately the pharmacology is too complex, and fixing this has not proven to be a fruitful approach to treatment.

The most fixable problem with depression seems to be that our natural servers delete e-mails at a rapid rate, sometimes so quickly that the message doesn't get through. This can be alleviated by preventing the decay of serotonin, or by preventing it from being reabsorbed by the neuron that was using it to send a message in the first place. Monoamine oxidase inhibitors (MAOIs) were the original antidepressants. They work by inhibiting the enzyme that degrades serotonin in the gaps between neurons. However, they also act on several other essential chemicals, so have a variety of nasty side effects, and can cause potentially fatal interactions with other drugs and foods.

They have generally been replaced by tricyclic antidepressants or by selective serotonin reuptake inhibitors (SSRIs). Prozac is the best known, but Zoloft and Paxil are heavily advertised, and there are at least a dozen others. Wellbutrin, preferred by many in part because it has less of an effect on weight gain and libido, inhibits reuptake of another signaling chemical, dopamine, but may act by indirectly stimulating serotonin release.

Each of these drugs has advantages and disadvantages, and it turns out that there is enormous variation in how people respond. Consequently, it can take months to years for clinical psychologists to find the combination and dosage that is right for any particular patient. Most settle on a cocktail that balances the improvement of mood with acceptable side effects, and once they have done so are advised to stick with the drugs for the rest of their life.

Of course, the major problem that plagues e-mail is spam. It is thus surprising that there is precious little written about the possibility that too much messaging can clog up the serotonin system. Manic phases are largely attributable to an excess of serotonin, as are hallucinogenic highs, but that is not the same thing as too much junk signal getting in the way of normal communications. The idea that depression arises not so much from a deficit of serotonin, but rather from the buildup of resistance to it (much as diabetes arises from resistance to insulin), is just starting to gain traction. It is a controversial hypothesis, but an attractive one with respect to understanding the evolution of sadness, as we shall see at the end of the chapter.

Something that is not at all controversial is the role of stress in promoting depression. Psychological stressors such as losing a family member, experiencing prolonged financial hardship, or coping with mental abuse are regarded by many as kindling for depression. They are the sparks that light the fire.

One of the major responses to stress is the activation of cortisol. Cortisol is also the hormone that gets us out of bed in the morning and does many other things such as suppressing the immune system (making us susceptible to infection in the days after a prolonged period of tension) and contributing to obesity by mobilizing blood glucose. Conventional wisdom is that cortisol levels shoot up in depressed patients and contribute to the mood change, either directly or by feeding back to serotonin. Not surprisingly, the pharmaceutical industry is hot on the trail of cortisol inhibitors.

In the meantime, we have Relacore, America's number 1 belly fat pill. According to the manufacturer's Web site, excess tummy fat is not a woman's fault, but rather arises from the harmful combination of everyday stress, overeating, and excess cortisol. The recommended solution is to take this natural mood enhancer and antistress pill that makes you feel better and lose the belly fat by inhibiting serotonin's natural antagonist. Like so many of the advertisements for antidepressants that show melancholy people taking the drug and soon thereafter hiking with the family through mountain meadows, playing fetch with the dog, and generally enjoying life, you have to wonder whether there aren't more effective ways of combating the hormones that keep us down—such as going for a hike or playing with the dog.

Or attending psychotherapy sessions. Data suggests that just taking antidepressants works only for about half the population. Similarly, talking with an expert is effective for no more than half of us. But combining the two seems to work well, helping probably at least three-quarters of the clinically depressed population to manage their disease.

misbehaving serotonin

Depression is one of the most genetic illnesses there is; yet paradoxically no one has yet found a gene for depression. Certainly not in the sense of "if you have this mutation, you will be bipolar," but also not even conclusively in the sense of "if you have this variant, you're strongly predisposed." There are some suggestions, as we're about to see, but they're at best of the "maybe, sometimes, depends on the circumstances" type—which actually makes them interesting.

Even though bipolar disorder afflicts only a few percent of the population, identical twins have about a 70 percent concordance rate. That's maybe a fiftyfold increase in susceptibility by sharing all your genes (and the womb and upbringing) with your twin. Even nonidentical twins have a 1 in 4 chance of bipolar disorder if their twin is afflicted, also heavily implicating the genome. It is similar for major unipolar depression, though perhaps with a slightly reduced genetic component.

If you're thinking that the genes that mediate serotonin signaling might be good places to look for an involvement in depression, you're not alone. The serotonin transporter gene, variously known as *5HTT*, *hSerT*, or *SLC6A4*, has been a particularly popular object of genetic studies for three reasons. First, SerT is the protein that Prozac, Zoloft, and company act on. Second, many of the studies do actually detect some tantalizing link between variation and mood. Third, a couple of unusual features of the gene affect how much it is used.

Basically, humans have a short and a long form at the front of the gene, and another section of variable length in the middle of it. The one that has been linked to depression is the first one, known, for good reasons that need not concern us, as the 5HTTLPR. About half of us have one copy of each allele, while one-quarter of us have two copies of the long form, and similarly one-quarter have two short forms. It turns out that people with the short form make less of the protein, which almost certainly affects the amount of serotonin that lingers in the little gaps between neurons.

SSRI antidepressants function by preventing 5HTT from transporting serotonin out of these gaps, so it stands to reason that the 5HTTLPR should affect the onset of depression. Well, sometimes it does, and sometimes it doesn't, and after pooling the results of dozens of studies together, maybe the short form increases your chances of having depression somewhat, but no more than 20 percent. It is a surprisingly small risk factor, and also surprising in that the less active form provides the risk—unless continued elevation of the chemical over time leads to serotonin resistance.

The story does not end there, though. There are a couple of interesting subplots. Because major depression so often leads to suicide, people have asked whether this polymorphism might be involved in the tendency to commit suicide. Believe it or not, suicide also has a large genetic component to it. Several studies have suggested that the short form has a highly significant impact on the incidence of violent suicide, but not of nonviolent suicide. It came as a surprise to me to learn that there is a difference, and the authors of the papers assume this is self-evident. Presumably gunshots, hanging, and jumping off a cliff are violent methods. The speculation is that 5HTTLPR is having an independent effect on the tendency to aggression and self-mutilation, which combined with depression, pushes someone toward violent attempts on their own life.

Dunedin is a city of just more than 100,000 people situated close to the tip of the south island of the beautiful country of New Zealand. It faces out into the great Southern Ocean, next stop Antarctica, and bears the brunt of its cool, damp winds that bring clouds, drizzle, and the sort of conditions that make a person sad. It seems an unlikely place to carry out a groundbreaking study of the genetics of depression, but nevertheless that was the case back in July 2003.

The research team followed almost 1,000 Dunediners from birth through young adulthood. They recorded the number of stressful life events each person had to cope with between the ages of 21 and 26: One-third of them were charmed; one half had one or two episodes of financial, health, relationship, or similar stressors; and the remainder were not so lucky. Around 17 percent experienced some measure of depression in their 26th year, and 3 percent reported either attempting suicide or thinking about it a lot. The startling finding in this study was that neither stress nor genotype alone had much of an effect on measures of depression, but the combination mattered a lot. Experiencing three or more stressors and having at least one copy of the short 5HTTLPR more than doubled the probability of depression symptoms, of clinical major depression, and of suicide thoughts. Additionally, adults with the short form were found to be far more likely to suffer episodes of major depression if they had been maltreated as children.

This was a landmark study because it confirmed with hard data the long-held suspicion that the interaction between a person's genetic makeup and their environment is what really matters. A dozen groups around the world have since had some success in replicating the finding, though it must be said that not all of them succeed. That is not really surprising, given the complexity of both the biology and of study design. Interestingly, the same group a little earlier had found a similar interaction between a gene that encodes the enzyme that degrades serotonin, monoamine oxidase, and the experience of parental abuse as children. Again, the type of allele or upbringing alone had little effect, but the combination of genetic and environmental risk factors increased the incidence of acts of violence and other antisocial tendencies committed by the teenagers and young adult men. These types of studies are notoriously controversial and the specific conclusions should be taken with a grain of salt, but they point in a direction that many believe must be true, namely that what genes do is very much a function of everything else going on with the person they find themselves in.

faint genetic signals

All this searching for the key to depression underneath the lamplight of serotonin is fine, but we need to remember that it takes a genome. When the book on the genetics of sadness is finally written, it will surely also include entries on the enzymes that synthesize the chemical and on the receptors that take up the signal. It is increasingly apparent, though, that it must include dozens if not hundreds of other chapters that can barely be sketched in outline yet.

The search continues using more traditional approaches. Linkage mapping in pairs of affected twins has turned up a dozen possibilities. This is where researchers look for parts of the genome shared by nonidentical twins who both have the disease. As a matter of logic, siblings always share about a half of their genes, but if the same region of a chromosome is shared in a sizeable fraction of hundreds of twins, it suggests that there is a gene thereabouts. One strong possibility is a gene called *PREP*, not because depression is a preppie disease, but rather because it encodes prolyl endopeptidase, an enzyme that processes a class of mood-altering hormones in the brain.

The contemporary genetic approach is to scan the entire genomes of thousands of unrelated patients for variant SNPs that associate with disease. This has worked quite well for the other diseases we've considered, but when the British Wellcome Trust spent a few million dollars on the venture last year, they came up with...nothing. Not even the *Serotonin transporter* gave more than a hint of a signal. It is a striking result.

Taken at face value, it means that there just aren't common genetic variants that consistently lead to chronic despair. There's a chance that they missed one or two, and maybe if they'd studied some group other than Englishmen they'd have had better luck. But more likely, the underlying assumption that common diseases are due to common variants is just wrong in this case—at least, not common variants that increase the risk by more than 20 percent.

There are a couple of alternatives. **One is that thousands of sites in the genome contribute, but each one has a barely measurable impact—depression by 1,000 genetic pinpricks.** Since morose, moody, or just plain contemplative people tend to end up marrying one another—just as tall people tend to hook up, and scientists pair with scientists, and alas white folk with white folk—these small effect variants would concentrate in families. So would the disease susceptibility, though it would be difficult to find the specific alleles involved.

The other possibility is that hundreds of genes have rare mutations that make a big contribution, but only in occasional cases. On this model, major depression would have a different genetic basis in different people, but whatever the issue is, it would lead to similar problems as far as signaling in the brain is concerned. These mutations would not have to affect everyone who carries them. They could individually increase the risk of depression twofold, or twentyfold for that matter, but would go undetected because only every ten thousandth person has any one of them.

schizophrenia and other mental disturbances

Iceland would seem to be another unlikely place to find genetic studies of mental health being carried out. Nestled up there in the middle of the North Atlantic Ocean, next stop the Arctic Circle, it is famous for icebergs, glaciers, and winter days that are about as long as a soccer game and doubtless just as dull. Nights are another matter since Reykjavik is also renowned for the beauty of its women, and quite the party culture has grown up there. For the past decade, Iceland has been home to the most ambitious, controversial, and in many respects productive, human genetics study on the planet.

The reason for this is that a prodigal son named Kari Stefansson returned from Harvard in 1996 to set up a pharmaceutical discovery company registered as deCODE Genetics. With the blessing of the government, and financial backing from Hoffmann-La Roche, the company gained sole access to the intensive medical records of most of the 270,000 citizens of Iceland, along with family pedigrees going back centuries to the Viking times. If you've ever read *Njal's Saga*, you will know that ancient Icelandic society was very much based on brutal retribution that makes an eye-for-an-eye justice look timid. Good family records were thus a must.

Although this right of access has since been overturned in court, the majority of the citizens have in fact granted not just medical records but also blood and hence DNA samples to deCODE. The combination is a rich treasure chest that human geneticists salivate over (while bioethicists make a living worrying about the consequences). It will really get interesting if and when the company starts churning out complete genome sequences for all of these people: Iceland may well be the first country to have the complete genome of its population determined. Out of this mix has emerged the discovery of novel genes for a bundle of common diseases, including prostate cancer, type 2 diabetes, glaucoma, psoriasis, asthma, osteoarthritis, stroke, cardiovascular disease, and even restless leg syndrome.

The reason I mention it here, though, is because deCODE is also honing in on several strong candidates for schizophrenia. Many of the things we've discussed about depression are equally true of this delusional psychosis. These include an incidence around one percent of the population, typical early adult onset, and presumed origins in complex genetics and environmental risk factors. Replace serotonin with dopamine, which is another neuronal signaling chemical, and the stories show distinct parallels.

The tale of the genetics of schizophrenia is also sprinkled with hopeful leads and heartbreaking dead-ends. Time and again a result that appears to explain some proportion of the inheritance of the disease has simply turned out to be a nonfactor in another population. This failure makes the experts wonder whether it is a factor in the initial study as well. Either the genes identified are terribly unreliable indicators, or they are specific for different ethnicities or cultures.

Nor does it stop there. Autism has also failed to yield to genetic dissection even though we know it has a heritable component. Mental retardation afflicts a few percent of all children, but with the exception of Fragile X syndrome where a particular gene is well implicated, it looks unlikely that common variants explain the disease. As a class, then, mental illness seems to be even more complicated than other diseases.

One reason for this may be because so much of the genome is used in the brain, well more than half of it. So there are thousands of genes that can go wrong, in each case predisposing to some sort of psychological disturbance. In fact, a couple weeks before this book goes to the printer, papers have started to appear that strongly implicate new mutations known as copy number variations in both autism and schizophrenia. It is beginning to look as though as many as ten percent of sporadic cases of these psychological diseases occur because children a born either with an extra copy, or missing a copy, of one of a handful of genes—so far— and that this disturbance alone deeply upsets the brain. That there aren't more of these psycho-CNV surprise us. It either is a minor miracle or can be thought of as a sign that evolution has found ways to suppress the effects of potentially harmful mutations.

the genetic tightrope of the mind

What possible good can come of genes that would plunge a person into such depths of darkness that he would want to take his own life? How can we understand why millions of housewives, having chosen to devote their best years to the hopeful raising of wonderful children, nevertheless begin each day with sheer dread at the thought of making breakfast and feel the terrible burden of sadness take over from there? Why do our genes let that happen? And why must genes be responsible for all the children who are so full of anxiety that they cannot converse with their parents, let alone the world at large? How is it that an evolutionary process that has given us Michelangelo and Beethoven, Einstein and Shakespeare, nevertheless endows so many of our most creative artists with the capacity for such debilitating depression?

Many would have you believe that there is some benefit to despair, that it is necessary for us to be able to appreciate great joy, or to be more prosaic that it is the price we pay for possessing a contemplative consciousness. In fact, it is by now almost obligatory for anyone who offers commentary on mental health, whether in a book, on Wikipedia, in a blog, or through a 2-minute television sound bite, to pass judgment on the adaptive value of melancholy. Yet we don't go out of our way to make such arguments to justify cancer, heart disease, asthma, and arthritis, so there is no reason to go that way for diseases of the mind. You should know that for every evolutionary theorist who promulgates these ideas, there are at least as many practicing evolutionary geneticists who daily get their hands dirty digging around the genome, who reject them as at best unnecessary, at worst just silliness.

For a different perspective, contemplate a tightrope walker. Maybe you have seen a funambulist at a weekend fair, just a couple of feet off the ground, hands held out for balance, or if they are really good perhaps even juggling some balls. There is not much risk, but a lot of joy, and a gentle metaphor for the capacity of the mind to assimilate all kinds of tendencies that would push us into a fall. Now picture one of those maniacs traversing a high wire strung between two tall buildings, supported by little more than a long balancing pole that shifts her center of gravity to a manageable place. This is higher consciousness, a tightrope act if ever there was one, capable yet manageable given the devices at hand, but always at risk of falling into an abyss. Now give her a pole from which you've docked a few feet from one end, without telling her, and ask her to cross the Niagara Falls on a windy day, ferocious cascades churning away underneath for dramatic effect. This is the human mind, a delicate balancing act weakened by genetic defects we know little about, trying to cope in a hostile new environment of our own making.

Talk about genetic imbalance and the origin of disease. The argument in a nutshell is that mood is such a complex trait that over millions of years mechanisms have evolved to buffer it from all sorts of stresses. Organisms must cope with a frightful environment, and mutations can hit thousands of genes that tend to disrupt the way the mind works. The human brain evolved in the genetic wink of an eye, fundamentally changing our mental capacities but leaving us more exposed, less buffered, than ever. On top of this, in the last few generations we have created a totally novel environment in which it is expected to act. We should be wondering why any of us are able to cope at all—but what's that old adage, all the world's a little strange, except for you and me, and even thee I have my doubts?

Consider first the environmental and cultural changes. No doubt that bangle of colored balls strung across a baby's crib, or that annoying toy piano that delivers an electronic mimic of some animal when she puts her finger on its picture, is stimulating enough. But it is not a real substitute for the visceral contact with a real world that every infant mammal has grown up with for the past hundred million years. Not that we should go back to hunter-gathering, growing up in tents, and moving our home every few weeks in search of food, but I am saying modern humanity is different from day one.

Growing older, we increasingly inhabit a virtual world. A friend of mine told me the other day that he found his precocious two-year old not just on the computer, but playing a flight simulator game online. We all lament that our kids are more adept at hitting a baseball out of the park on a PlayStation (or, more recently, a Wii station) than they are at running the bases around the cul-de-sac at the end of the street, or that they prefer piling up thousands of hours of cell phone bills to reading 100 pages of a good book, but do little about it. Then there is the violence, sexual innuendo, and nonstop banality of television, and the in your face propaganda of talk radio, and we wonder why adolescents have a hard time adjusting to the modern world.

On the other hand, it never ceases to amaze me how engaged and energetic college students are. The party-versity that Charlotte Simmons finds in Tom Wolfe's novel is real, but so too are the many twenty-some-things who are putting themselves through school, studying a full 40 hours a week, putting in another 20 hours of community service, and somehow finding time to play a sport and lead a full and healthy social life. Those who aren't in college are as often as not working two or three jobs just to pay the rent and car loan while keeping the credit card debt somewhat manageable. It is the pace of life and the enormity of the pressures this all places on us that is so different. Humans have always dealt with adversity in their daily lives, but the grind today is nothing like anything ever before. Only in the past century has every single one of us been expected to be active from dawn to midnight with so little downtime and so much peer pressure to keep up, lest we be socially outcast or thrown on the junk heap of a failed career.

Then there is the stress on relationships. Take a walk through the airport terminals in Chicago, Los Angeles, or New York and take in the faces of the travelers. Billy Joel's scene at the bar has become the scene at the Chili'sToo as waistlines grow ever larger and mascara runs into the stress lines of harried faces. Pursuit of the golden buck that will keep our loved ones close and happy drives businessmen to the place they call loneliness, and young people's pursuit of a more exciting and meaningful life drives a wedge into the heart of family. Facebook and MySpace might keep acquaintances in touch, and surely we've just scratched the surface of virtual friendship, but the mere fact of their popularity says something about how drastically human interactions have changed. It has to be affecting the way the mind develops and carries on.

Finally for now, but certainly not exhaustively, there is the stress of modernity. Martha Stout speaks of "limbic wars" in her book, *The Paranoia Switch*, arguing that Americans are so fearful in the wake of the 9/11 terrorist attacks that an almost contagious rewiring of the neural circuitry has occurred that has reshaped the way we perceive one another. By undermining our sense of tolerance and compassion, the terrorists have hurt far more people than they ever could place in direct danger. She further points out that there are maybe 1,000 times as many women suffering today in physically abusive relationships than were killed in the attacks. Marital warfare takes a far stronger psychological toll, yet the nature of it is similar to the imprint of three generations of almost constant exposure to military warfare. World War II, the Vietnam War, the Cold War, and now the War on Terror have surely placed a stress on our collective mental health.

So much for environmental imbalance. What of genetic imbalance? *Homo sapiens* is by any measure one of the younger species on the planet, having been around for fewer than 10,000 generations. By comparison with the average species, we can expect to be around for at least a 100 times that long before we evolve into something else or become extinct. This is also about the length of time since we shared a common ancestor with our closest relatives, the chimpanzees.

A lot has happened in our short time on the planet. Using new statistical methods focusing on those parts of genes that regulate when and where they are used, we can see that genes involved in brain development seem to have been particularly subject to change. Maybe as many as a quarter of our brain genes are used ever-so-slightly differently in humans, though the liver and testes are at least as divergent. It is hard to know which or how many specific changes have been key to the emergence of intelligence, and which have gone along for the ride.

Actually, we don't need to look at the genes to conclude that there is something different about the human brain. Our reasoning abilities are unparalleled, at least on this planet. We communicate with symbolic language that has just bare rudiments in other primates, and we use signs and tools to build cultural practices that have as profound an impact on humanity as any of our unique genetic attributes. We try to engage in monogamous relationships, for the most part, and live in extended family relationships that are definitively human, typically extending the trust we share with family to friends as well. We have a spirituality that is undetectable in other creatures but seems to be at the very core of what it is to be human, though it is often allied with a sense of wonder that can create a sense of emptiness that is at the heart of depression.

a kindling theory in the modern world

The final piece of the puzzle is to realize that the mechanisms that regulate mood have been exquisitely fine-tuned and genetically buffered, or canalized, ever since the dawn of animals. All the major brain signaling mechanisms, including serotonin, dopamine, and a half dozen others, are in place and working pretty much the same way in flies, pythons, and piranhas. I'm not sure what a depressed python or a manic piranha looks like, but just like us, their mood is regulated within strict bounds that enable the animal to function, and it would not surprise me if some of these animals also have mental health problems.

The buffering is the key here. Just like blood glucose levels that rise and fall with the day and with the stages of your life, serotonin levels adjust to a person's circumstances. Whether you are fighting or fleeing, sleeping or waking, angry or amorous, comes down at some level to this chemical. Despite the wide daily variation in activity, it is crucial that the basal levels remain within certain bounds, lest your whole mood system gets out of whack.

This is why addictive drugs are so dangerous over the long haul. Cocaine inhibits the reuptake of dopamine, nicotine prevents another signal called acetylcholine from finding its receptors, and hallucinogens mess with your serotonin function. The brain responds to all these drugs by modulating the levels of the receptors: Constantly firing in a bath of elevated serotonin, the logical response of a neuron would be to desensitize itself by turning down the number of receptors. This is what it does, but it means that without the drug an addict is left desperate for more of the signal to overcome the lack of reception. Over time levels can return to normal, but often, heavy drug users are left with a state of chronic low-level depression.

Another sign that the brain modulates itself is seen in the phenomena of sensitization. A fly that is fed volatile cocaine will go a little crazy, buzzing all over the place. As the procedure is repeated, it becomes progressively wilder, crashing about its home like a pinball. Stimulate a rat's brain with electric shocks at low intensity, and after a couple of weeks it starts convulsing, having become more sensitive to the shocks. In humans, over time, the frequency and intensity of epileptic seizures generally increases, and lately many physicians have begun to wonder whether the same is true of bipolar disorder.

The hypothesis is called the *kindling theory*. Kindling is the dried small sticks and twigs that we used to use in the days before MatchLight inflammable fluid-soaked charcoal meant that fires could be guaranteed at the strike of a match. So perhaps we should rename this the Match-Light theory, reflecting the idea that we're all primed for mental disturbance in today's fast-paced world. In any case, the idea is that once one episode of depression occurs, the changes in the circuitry between the neurons that ensue leave a person prone to another episode.

It is also why the notion of serotonin resistance is so appealing as a causal factor in mood disorders. Over hundreds of millions of years, regulatory systems evolved to ensure that highly reactive signaling mechanisms in the brain stay operational within safe bounds. They include ways to fine-tune the volume controls in the limbic system, but the extraordinary changes in human history have pushed these buffering mechanisms to the limit. The ancient limbic system that regulates mood now has to integrate high-level brain functions that have been around for just hundreds of thousands of years, not to mention cultural practices that have been around for just tens of thousands of days.

As a result, we're hypersensitive to new mutations that naturally appear in all those mood-related genes every generation. A person who happens to inherit several of these, no matter how rare they are, is at risk.

It may take a few thousand generations to recalibrate. Or it may be that selection in modern humans is so weak that there never will be an opportunity to recalibrate and those little mutations will gradually accumulate in the gene pool. There's a depressing note to end this chapter on: Maybe the ultimate fate of the human species is never-ending deterioration of our mental stability.

7

The alzheimer's generation

slow walk to dementia   Alzheimer's disease (AD) is the gradual unraveling of memory in a virtual reversal of the lifetime assembly of the mind.

alzheimer's on the march   By the age of 65, less than one percent have AD. By the age of 85, as many as half of us will have it. Without a cure, the Baby Boomer Generation may eventually go down in history as the Alzheimer's Generation.

tangles and plaques   Tangles of Tao and plaques of Amyloid beta clog up the brain, but it is likely that these good proteins turn bad long before autopsy confirms their guilt.

early onset FAD   Familial AD is for those in their forties and is traced on the whole to just three genes, the building block and scissors that make Amyloid beta. If it is in the family, perhaps it is best to see a genetic counselor.

late-onset LOAD   The vast majority of AD is late onset, beyond retirement. There is one major risk factor, *ApoE*, but even that is far from diagnostic. The risk comes from our primate ancestors, and most people are actually protected by the modern allele.

just growing old   Extension of life span is one of the most remarkable of all human attributes, and the last two generations have seen the most dramatic gains in the history of life on Earth. No surprise then that we're out of equilibrium, and that the most complex of our organs, the brain, often gives way first.

slow walk to dementia

Surely the most aggravating sign of the teasing and fickle nature of our genes is their insistence on gradual decay. And there is no more certain sign of their power than the decline of Ronald Reagan. Within a decade of leaving the Oval Office of the White House, the man who helped bring the Soviet Union to its knees was himself reduced to a mental crawl by a handful of errant nucleotides sitting around in his genome.

Many would say they saw the signs of Alzheimer's disease well before he even left the Presidency. But who can say whether his inability to remember the press corps' questions was an early sign of dementia, or the polished acting performance of a master politician? What we do know is that the course of the disease was little different from that which tens of millions of Americans now face. The gradual unraveling of the mind, the journey from maturity through despair to senility, unfolds as a mirror image of the 20-year journey from infancy through hope to adulthood.

Nothing generalizes anxiety more than early signs of memory loss in middle age: misplacing the car keys on a regular basis, or forgetting the name of a casual co-worker at some embarrassing juncture. **If someone in your family is known to have suffered from Alzheimer's, perhaps only a great-uncle you never met, the natural presumption is that you are next in line, that the genes that somehow bypassed your parents have caught up with you and now sentence you to a decade of dementia**—inevitably to the point where not only can you not recall a conversation from the day before, but you do not even recognize your own self.

The fear is enhanced by the knowledge that close friends and family will bare the brunt of the burden. They will be the targets of your cantankerous acts of cruelty; they will take you to the bathroom and clean up the filth; they will daily live the hell that after a time becomes oblivious to you. They will be the ones who struggle after you are gone to displace the memories of your failing days with the positive images of the real person who accomplished so much, gave such love, and meant the world to so many.

Then there is the financial worry. The majority of health care is spent on the last few years of life ordinarily, but with Alzheimer's patients the costs are spread over a decade, many not covered by insurance. Nursing home care and medical expenses are estimated to start at $70,000 a year, considerably more than Social Security will cover, so they eat into savings meant to provide a long and happy retirement for your spouse, if not a nest egg for the kids. This equation doesn't factor in the lost earnings of the daughter who gives up a couple of years and possibly a promising career to provide personal care.

Adding further to the anguish is the inevitability of it all. Yes, eating more fish rich in omega-3 fatty acids helps ward off memory loss, ginkgo biloba seduces many into thinking it does, doing crossword and sudoku puzzles can't hurt, and an active mental life is perhaps the best advice of all. A few years ago, Elan Pharmaceuticals grabbed headlines with a vaccine that showed enormous promise in halting the course of disease, until a fraction of patients came down with pathological inflammation of the brain. No one knows when a viable cure will emerge, so for now diagnosis of dementia is like a slow death sentence with little hope of reversal. May we each have the grace to deal with the progression with humor and understanding.

alzheimer's on the march

The generation affectionately known as the Baby Boomers may yet go down in history as Generation A, the top of the letter A standing for that great pyramid of aging boomers destined to suffer more from Alzheimer's disease than any generation in history. The statistics are stark: more than 65,000 deaths a year, five million Americans with Alzheimer's currently, that number set to nearly quadruple by midcentury and to be dwarfed by the worldwide burden in the year 2100.

The probability of contracting Alzheimer's doubles every three years after the age of 65. At retirement, only one percent of people are afflicted, but five percent have it in their early seventies, the fraction rises to one-quarter by the early eighties, and at least one half of all nonagenarians have AD. I have looked high and wide in an effort to ascertain whether this represents an epidemic independent of people just getting older, but there does not seem to be any clarity on the issue.

You would think that since Dr. Alois Alzheimer described the first case only in 1901 there must have been a dramatic increase in incidence of this form of senile dementia in the past century. His patient, Auguste Deter, was a woman in her early fifties living in Frankfurt, Germany, who presented to a psychiatric clinic with symptoms of memory loss, delusions, and mindless episodic behavior. She died within five years, not just the first recorded case of Alzheimer's disease, but also the first case of the early onset form. The good doctor was a colleague of Emil Krae-pelin, the man widely regarded as the father of the view that psychological problems generally have an organic biological, rather than purely psychological, basis. Together they performed an astonishing series of histological examinations on Frau Deter's brain after her death, describing the characteristic cellular anomalies that remain the only clear definition of the disease.

This is part of the reason why it is so difficult to know whether AD is on the rise. Senile dementia has been widely recognized throughout recorded history, but until the last century was more likely seen as a normal phase of life than a disease that afflicts some people and not others. Only in the past decade or so has it been widely appreciated that around two-thirds of cases of acute dementia are actually this genetically influenced disease. Without autopsy, there is no sure diagnosis, the major alternative explanation being vascular dementia: basically, the consequences of myriad small strokes that increase in frequency as a person ages, slowly degrading the brain tissue. To this day, I have no idea whether my own grandmother had AD. She certainly lost her memory, went through a cantankerous period, and had odd behavioral lapses in her nineties, but no diagnosis was ever made that we are aware of.

It is well known that the average age of people has increased over the past century. In North America, Europe, and Australia, more people are in their late thirties than any other age group, and most of these can expect to live well into retirement. China is just a decade behind, South America two decades, and India will get there by the middle of the century. For this reason alone, the absolute number of Alzheimer's patients is set to rise toward the billions worldwide as the century proceeds, as will the proportionate contribution to human morbidity.

The key statistic, though, is whether the fraction of people in each old age group with Alzheimer's is increasing. It probably is, suggesting that any genetic variants that predispose to Alzheimer's are becoming more potent. Some would argue that really what is happening is that the proportion of older people who have chronic health problems is increasing, particularly those with diabetes or coronary heart disease. Companies that make daily pill boxes would seem to be just as good an investment as ones that make baby clothes. Since Alzheimer's might well be correlated with these metabolic diseases, the at-risk population is growing as a proportion of the elderly, who used to be an unusually spritely bunch. Even so, we're almost certainly once more seeing an effect of the modern environment interacting with the historical human genetic endowment to produce a common disease.

tangles and plaques

When I went to take my dogs for their walk this morning, Razzie's leash had a knot in it. No big deal, it only took a minute to undo, though a minute is a long time for a puppy with a world outside to explore. The weird thing is that I have no idea how this happened, in the same way that it defies common sense for the garden hose to get itself all twisted precisely and only on those days when the far back corner of the lawn needs watering. It is clearly in the nature of things to get all tangled up.

This is also why fishermen spend half their lives on dry land, repairing their nets. No matter what they trawl through, nets pick up all sorts of detritus and gunk that clogs things up, and attract ribbons of seaweed that weave their way around the strands. In time these begin to eat away at the net, leaving holes and leaky gaps, and left untended, before long the whole thing is useless.

The brain too is a network, of axons and synapses, sadly vulnerable to decay as it trawls through oceans of thoughts. **When Alois Alzheimer stained Auguste Deter's brain, he found it to be gummed up with plaques and tangles, and this pathology turns out to be the hallmark of the dementia that takes his name.** All that white and gray matter may seem random and disordered to the untrained eye, but it is actually highly organized, with layers of complexity seemingly added according to experience. As a child grows, the neurons tie together, searing memories into the very structure of the assembled synapses. As the plaques and tangles grow denser, they undo these memories and seed degeneration, and the patient's thoughts regress in an orderly reversal of their maturation.

The stages of undoing eerily mirror their creation. First the ability to perform professionally and complete complex tasks goes, followed by simpler daily chores such as cooking and organizing schedules and personal finances. Then the things we take for granted such as getting dressed and taking baths are no longer possible. Incontinence follows, along with the inability to speak and even express emotions with a smile, all because some people are prone to growing plaques and tangles as they pass retirement age.

We know what they are and where they come from—they are made from our own genes—but have little idea yet why the acute accumulation occurs. The plaques are made of a jumble of protein pieces called Amyloid beta, or Aβ, peptides. These are cleaved from a longer protein known appropriately enough as Amyloid Precursor Protein (APP) that is thought to be required for normal brain development. The Aβ pieces probably don't do much harm just floating around neurons on their own, but they can change their conformation and start aggregating into fibrils. When these are deposited outside the neurons they assemble into the characteristic Alzheimer's plaques.

Meanwhile, another essential protein, *Tao*, also aggregates inside the neurons, forming the so-called neurofibrillary tangles. *Tao's* normal function is to stabilize the framework of the neurons: All cells have a kind of dynamic scaffold that keeps getting turned over, and *Tao* has an important and active role in this. If it becomes overactive, it assembles into long strands instead, creating the visible tangles.

Two questions arise immediately. First, what is the relationship between Aβ, Tao, and dementia; and second, why does it take six decades for them to become pathological? I hope by now that you are not expecting a straight answer.

For one thing, it is not at all clear any more that the plaques and tangles aren't themselves just symptoms of whatever is going wrong. Toxic waste dumps are bad enough, but are just as much a sign that something else is going on that needs attending to. A growing body of evidence suggests that it is not so much the large plaques of Aβ aggregation that crop up between the brain cells, as it is the couplets or small gangs of these little protein pieces that are doing their damage inside cells. In the case of Tao, surely the long tangles aren't much good for a person either, but they twist together only when they have gone absent without leave from their normal jobs. It also seems that they have a knack for making a royal pain of themselves, by glomming on to all sorts of proteins inside cells, including Aβ.

The second question is even trickier, and I'd be lying if I implied that anyone has a good explanation for why proteins that work fine for 60 years suddenly turn bad. No one really knows why some of our finest writers remain as sharp and insightful into their nineties as they were in their prime, while others—Iris Murdoch and Ralph Waldo Emerson among them—are destined to lose their minds as surely as anyone else with AD. Something is disrupting a lifelong equilibrium between active forms of these proteins and their partners, setting off a chain of events that cascades into disease. It may have something to do with the oxidation and structural damage that nags away at us throughout our earthly time. But even this begs the question: Why some individuals but not others?

early onset FAD

A century after the disease was first described, we now have a good sense of the genetic factors leading to the early onset form that runs strongly in families. Three genes have been implicated: the precursor of Aβ and a pair of scissors that release it. These are thankfully rare forms of AD, but there are now more than 300 families in which three generations of early onset Alzheimer's sufferers have been found. Most of these cases are attributable to one of more than 150 different mutations in the three genes. Some cases also clearly run in families but cannot be ascribed to these genes. For the most part, though, if there is a cluster of relatives all of whom show signs of dementia in their mid- to late forties, geneticists are likely to be able to devise a test that will tell younger family members whether they are likely to suffer as well.

The most common culprit is called *PRESENILIN-1*, or *PSEN1*. Responsible for more than half of all early onset familial Alzheimer's disease (FAD), it works in conjunction with the second culprit, *PSEN2*, to cut Aβ out of the APP precursor protein. Together they are part of the gamma secretase complex, not to be confused with the alpha and beta secretases. All these protein scissors have natural roles in appropriately cutting up APP (and various other proteins that need not concern us here, but feel free to look up *Notch* on the Internet). Alpha cuts off a bigger piece of APP that includes the start of Aβ, but it competes with beta for APP's affections. The combination of beta with gamma ultimately drives up the levels of Aβ inside and outside neurons.

Then there are the *APP* mutations, at least 15 known so far. Most of these change the genetic code exactly at the part of the protein that gamma secretase cleaves. One allele found in Sweden increases the ratio of beta to alpha secretase activity, others just increase them all, and some may even have the opposite effect. Ultimately, they all increase the amount of Aβ, or more precisely of an ever-so-slightly longer form of the protein that is just a little bit more pathological. All this stands to reason and confirms the causal role of this protein in promoting Alzheimer's disease.

APP is also the reason why adults with Down syndrome invariably also suffer from early onset Alzheimer's. The gene is on chromosome 21, which is present in three copies, instead of the usual two, in Down's people. Presumably with an extra copy of the gene, more APP is produced by the neurons, and this upsets the regular balance that the secretases deal with, resulting in prematurely high levels of Aβ. Most of the other Down's features are thought to similarly arise because of an imbalance in the number of copies of one or more of the several hundred genes on chromosome 21. All but one of our other chromosomes are considerably bigger, and if a fetus has three copies of one of them by accident, it is likely to result in spontaneous abortion before too much development proceeds.

late onset LOAD

Late onset Alzheimer's disease (LOAD) is a completely different story. It is more like all the other complex diseases we've considered so far. **Although 50 percent of people with an affected brother, sister, or parent can expect to get Alzheimer's by the time they reach their eighties, this percentage is not that different from the general population.** It is natural to worry about your own fate when caring for a loved one, but the fact is that genetic counselors can tell you little about whether you are going to have senile dementia. On the other hand, if both sides of your family are clear, it is a good bet that you will be okay; but then it is better to look at it as though you are lucky enough to have an unusually protective genetic constitution.

There is one major susceptibility factor, and that is a gene called *ApoE*. It is the only gene known beyond a shadow of a doubt to affect late onset Alzheimer's disease. Your personal risk changes more than twentyfold, depending on which alleles you have. But even this is not sufficient to predict the course of disease, so physicians are loathe to even try to do so. In fact, *ApoE* probably has more of an influence on age of onset than whether you will get Alzheimer's disease at all—namely, whether it is likely to come on in your sixties, seventies, or eighties.

ApoE is short for apolipoprotein E. That's about where I turned off in college biochemistry as well; suffice it to say that it has an important role in what happens to your cholesterol. That it is a major player in the etiology of Alzheimer's disease came as a major surprise, since until this discovery it was being investigated for its role in clogging up the arteries. Some investigators suggest that it is also involved in protection from viruses and possibly impacts fertility as well. All these functions are likely to have shaped the distribution of variation considerably more than the gene's role in dementia late in life.

There are three common alleles, which we will call °E2, °E3, and °E4. If you had your druthers, you'd ask to have two copies of the °E2 allele, please, as it seems to be associated with lower incidence of heart disease as well as AD. However, few people are so lucky. Worldwide, between one half and two-thirds of us instead have two copies of the °E3 allele, which is fine. It seems to be neutral. Depending on where you live, with no easily described global patterns, between 10 and 20 percent of us have at least one copy of °E4, and just a few percent have two copies of it. For each additional copy of °E4, your chance of getting Alzheimer's increases enough to make you worry if you start forgetting your keys, but not enough to be sure that it's going to be a problem.

Now, if you're wondering why anyone would have °E4 if it is so closely associated with disease, think again. It turns out that °E4 is the ancestral allele: Chimpanzees have it, and °E2 and °E3 are on their way to becoming the standard human condition. So, if anything, we are evolving resistance to Alzheimer's disease, though with some setbacks in places such as Scandinavia, South Africa, and New Guinea where °E4 remains common. Across central Africa, more than a quarter of the alleles are °E4, but the good news there is that early reports suggest it is not associated with Alzheimer's. If substantiated and combined with the fact that °E4 does increase LOAD susceptibility in African Americans, these results imply that our modern diet and lifestyle are interacting with ApoE to elevate the risk.

If you're wondering what a protein that binds fats for a living is doing in the brain, the answer seems to be that it is involved in the cleavage or transport of APP into or out of neurons. In other words, it is not exactly clear, but gamma secretase makes Aβ at the surface of cells, right in the midst of the lipids that form the walls of the neurons, in a place where ApoE likes to spend some of its time. So it seems to have a direct role in pathogenesis, in addition to any indirect linkage between heart disease and aging.

just growing old

Why do we grow old in general? For much the same reason that blue jeans and bicycles are disposable: It is a corrosive world out there. In time, chromosomes fray like downtrodden cuffs, and proteins oxidize like the metal of handlebars and spokes. But this facile answer belies the existence of extraordinary variability in the rate at which organisms age, much of this demonstrably genetic in origin.

As a species, humans are remarkably well off as far as aging is concerned. Our 80 years double the life span of our closest relative, the chimpanzee, and among mammals only elephants come close. That alligators are granted almost 70 years on Earth seems unjust, but swans make up for this by exceeding 100 years, and if you want a companion for life, then getting a parrot the day you are born is not a bad idea. Among dogs, small breeds such as terriers and miniatures typically surpass 14 years, but larger ones and bulldogs are lucky to make seven. In all these species, aging appears to follow a similar course, even in mice that live a scant four years.

We know that genes affect how long we live, because of mutant strains of worms, flies, and yeast that live much longer than usual. Remarkably, a common set of genes keeps emerging from studies of these model organisms, and it is widely believed that the biochemistry of aging is similar in mammals. One pathway is related to insulin signaling, which is best kept as low as possible throughout life. Similarly, and probably not coincidentally, dietary restriction (in essence, reducing how much you eat) is the most sure-fire way to live a long and healthy life. Another reduces the damage caused by "free radicals" generated by cellular energy factories over a lifetime: superoxide dismutase, or SOD, is basically rust proofing for the inside of cells. Then there are the Sirtuins, whose function remains obscure, except that they are the main target of resveratrol, the ingredient in red wine (and lately dietary supplements) that is supposed to make a glass of Shiraz each evening good for your health.

Step back far enough, and each of these pathways converges on a key gene called *FOXO*, which we can think of as the foreman of stress. When *FOXO* is active, it ensures that all kinds of stress responses are active. Heat shock, oxidation, xenobiotics, microbial immunity, DNA damage, metabolic dysfunction—protection from each of these keeps us from growing old prematurely. Some members of something called the Hormesis Society even go so far as to advocate taking a little bit of low-dose poison every now and then to keep *FOXO* on its toes, as it were.

Well, people who live long lives aren't mutant in the same way that long-lived *methuselah* flies, *age-1* worms, or *Sir2* yeast are. It is likely, though, that variation in the hundreds if not thousands of genes that engage in these types of activity is contributing to variation in human longevity. Further, they each interact in a small way with *ApoE* to influence the onset of Alzheimer's disease too. Crucially, different organ systems—muscle, liver, kidney, brain—all age at slightly different rates, in part because of the way they are wired genetically. Alzheimer's is premature aging of the brain relative to the rest of the body, in the same way that muscle wasting and heart disease are premature aging of those tissues. People get old like the houses they live in, bit by bit.

We may never identify precisely which sequences of the DNA code are generically involved in aging, because their effects are too small, particularly since chance also plays such a huge role in aging. Some will be common variants, some rare; some will be prevalent in Sardinia and others in Alaska; some will be worse for men than women. Most will be present in the gene pool for reasons that have little to do with determining how long we live.

Evolutionary biologists like to theorize about aging as a necessary counterpoint to having babies. They do so because the existence of genetic variation for longevity as well as profound differences in life span among species implies that the rate at which organisms age is affected by evolution, and by the strategies adopted to ensure survival in different environments. Some species are so extreme about this that they basically all die immediately after spawning or otherwise seeding: Bamboo plants live for 100 years, and then commit suicide as soon as they flower. This trade-off between reproduction and existence also enters into the human equation: Having many children early is not generally conducive to a long life for their mother.

These days, two types of evolutionary theory predominate. The "don't worry about it" theory, technically known as *mutation accumulation*, hypothesizes that mutations that act late in life have less of an effect on reproduction, so are more likely to build up in the gene pool. The "pay later" theory, technically known as *antagonistic pleiotropy*, hypothesizes that there are trade-offs between the good that mutations do early and the bad that mutations do late in life. Some aging-promoting mutations are actually thought to be beneficial to the species. The theories are not mutually exclusive, and both mechanisms are probably in play.

Both are predicated on the idea that natural selection favors alleles transmitted from generation to generation. Since the likelihood of death by predation and accident increases with age, a premium is placed on having children early, at least so long as those children can be raised successfully. In the don't-worry-about-it scenario, mutations that don't have an effect until after reproduction aren't going to have as big an effect on genetic fitness as ones that affect a younger organism. We have already seen that more mutations happen every generation than evolution can possibly cope with, so it follows that deleterious alleles that act late in life are not weeded out as efficiently as ones that affect children. Natural selection literally doesn't worry about them as much as it does about variation that acts early in life. Consequently, Alzheimer's is more common than progeria, that nasty disease where 12-year-olds are physiologically in their seventies.

The pay-later model differs in that it postulates that there will sometimes be an advantage to mutations that are harmful late in life—not because there is anything to be gained from killing off the elderly, but because these mutations do other things that increase the survival and fertility of the young. For example, accelerating the onset of menstruation allows for earlier childbirth, but increases lifetime estrogen levels and hence the risk of breast cancer.

To be honest, I am not just on the fence about which of these theories is more plausible, but actually somewhat skeptical of their relevance to the human condition. Modern humans live on average five times longer than they have throughout most of the history of the species. Paleontologists estimate that fewer than half of all children in the Old Stone Age lived past age 10, and life expectancy even in the Middle Ages was less than 30. Just in the past few generations substantial decreases in child and adult mortality have occurred, and immigrants to developed countries can generally expect to live much longer than their parents.

So the conditions that pertain to each of the other common diseases that we have considered also exist here. *Homo sapiens* don't reproduce until they are twice as old as other higher primates, and they are capable of living twice as long. Clearly there has been dramatic evolution of our genetic capacity for long life, too rapid for us to expect that we are at any sort of equilibrium. Correspondingly, cultural practices have dramatically extended human life span very recently. Together these factors ensure that a large fraction of those who reach their mid-sixties are no longer sheltered from the ravages of dementia.

8

Genetic normality

height and weight How tall and heavy we are is a function of thousands of genes, each with polymorphisms that have very small effects.

pigmentation A handful of genes suggest red hair and blue eyes, but on the whole human pigmentation is as genetically complex as any other trait.

the God gene Be wary of suggestions that God is in your genes, but don't be surprised if spirituality is.

a few words about IQ Why it is wrong to assume that genetic variation within groups implies that genes guide differences between them.

on being human Selection and drift have shaped the evolution of the entire human genome, and it is an ongoing process.

the adolescent genome revisited Fact and fiction in understanding the origins of disease.

height and weight

Imagine that your child is asked to bring pictures of a half dozen athletes to school for a show-and-tell on the theme of human diversity. Whom would she choose? Shaquille O'Neal and Nadia Comaneci perhaps: a behemoth of a basketball center at 7 feet, 325 pounds of pure athleticism, and the diminutive gymnast who achieved perfection on the uneven bars at the Montreal Olympics. An equally odd couple would be Takanohana Koji, one of the great sumo wrestlers of our time, and Maria

Sharapova, Wimbledon champion with the looks of a supermodel. Throw in El Guerrouj, the great Moroccan middle-distance runner, and the Aboriginal quarter-miler Kathy Freeman, and you have a pretty full coverage of human variation. We're not quite as diverse as Chihuahuas and Bernese Mountain Dogs, or Brussels sprouts and broccoli for that matter, but our genetic potential is unarguably diverse.

Another perspective on this is provided by a marvelous YouTube video clip, "Women in Film." As photographs of 77 Hollywood divas morph into one another—Bette Davis into Greta Garbo, Nicole Kidman into Catherine Zeta-Jones—you simultaneously feel how different and how similar people's faces are. It really is amazing how sometimes two completely different people can look alike. The guy who bought your old bicycle could have been the kid across the street when you were growing up, even though one is Asian and the other of American Indian extraction. Apparently Charlie Chaplin once entered a Charlie Chaplin look-alike contest and came in third!

What do our genes have to do with normal human variation, and do they tell us anything about the origins of disease?

Let's start with height, because this is possibly the most heritable of all human attributes and one of the most extensively studied. After factoring in generational effects, the correlation between children and their parents for height is around 80 percent (though whenever I try to demonstrate this in class the exercise invariably fails for some reason, presumably because we're bad judges of our own nature). By now, 40,000 Englishmen have had their DNA subjected to whole genome genotyping for one reason or another. That's plenty enough data to expect that scanning 500,000 genetic polymorphisms will lead geneticists to all the major genes that influence how tall a person is—except that it has only finger pointed 25 genes, and astonishingly they collectively explain a measly four percent of the variation.

Any one genetic variant is responsible for adding no more than a millimeter or two to height. Two people with the opposite alleles at most of these genes are certainly likely to differ by several inches in height, but they probably won't be in the 5-foot and 7-foot categories. If all these genes were identical in all people, we'd barely notice a difference in the spectrum from short to tall. These data also tell us that if we could measure everyone, we'd find that literally thousands of genes affect how tall we are. Some will affect growth of your leg bones, some the muscular support of the spine, and some how big your head is, but all the common variants have very small effects.

On the other hand, when small and large breeds of dogs are compared, it turns out that a few genes seem to account for many of the differences. Unbeknownst to them, those responsible for breeding most of the small dogs were selecting bitches and studs that have in common a single mutation in the *Insulin like growth factor 1 (IGF1)* gene when they chose which smaller than average dogs to mate. When it comes to explaining just the short legs of Dachshunds and Corgis and the like, a similar mutation in another gene seems to be responsible. There is even speculation that complex behaviors such as loving to play in water might have relatively simple genetic underpinnings.

How can we reconcile these human and canine findings? It is probably a reflection of the histories of the two species: Artificial selection by dog breeders acts primarily on mutations that have a large effect, while natural selection and genetic drift during human evolution has acted on background variation. Within single families, it is possible that there are alleles that ensure that siblings, even nonidentical twins, can be several inches different in height despite being raised in the same home. But these are rare enough in the general population that they don't make much of a contribution overall. Reciprocally, within breeds of dogs, there are thousands of polymorphisms that we will never know about that contribute to size differences.

The story for body weight is similar. Those tens of thousands of genome scans, many in the context of diabetes susceptibility, have pulled out just a couple of genes that have a relatively large effect. The major body mass gene in Caucasians is called *FTO*, which rumor has it is a contraction of *fatso* (but only fly geneticists can get away with such disrespect). We don't know what *FTO* does yet, but if you have two copies of the heavy allele, you may be as much as 2 or 3 pounds heavier than if you had the other alleles. Most of the remaining variation is due to literally hundreds of genes with alleles that add only a few grams here and there.

pigmentation

On the whole, humans are not a particularly colorful species, at least not naturally. Until a few decades ago, pretty much anywhere you went in the world, most people you encountered would have been the same local hue, somewhere along the bland axis from black to white with a touch of rouge or ochre thrown in. Ninety percent of the world's population has pitch-black hair, with only northern European derivatives experimenting with brown or blonde. Eye color is a little more variable, for reasons that are completely obscure.

Given this, Scandinavia would seem to be a good place to begin the search for the genes behind differences in skin color, and indeed the folks at deCODE Genetics in Iceland have obliged. Their scan of around 7,000 Icelanders and Dutchmen turned up five genes that convincingly influence eye and hair color and another region of a chromosome that is associated with freckles. Several of these were previously known to influence the transition from dark to pale skin in Europeans and East Asians as well, which makes sense because all these color attributes involve the deposition of a pigment in particular cells.

The biggest factor they found is a site near a gene called *OCA2* that increases your odds of having blue instead of brown eyes about twentyfold, and blue instead of green eyes about fivefold. More than 80 percent of northern Europeans have it. Very few sites elsewhere in the genome are more different in frequency between Africans and Europeans. No other genes come close with respect to their effect on brown eyes, but if you also have a particular variant of *SLC24A4* you can be reasonably confident that your eyes will be blue not green. It is not clear where hazel and gray eyes fit into the picture. Note that these are still by no means diagnostic indicators of color, which I find comforting, since my grandson has very blue eyes—even though his mother is without doubt an Italian woman. Were the dark brown eyes my stepson's instead, I suspect we'd be wondering a little harder about his role in the conception.

Then there is the melanocortin receptor, *MC1R*, which has two variants that pretty much guarantee red hair. Both are rare, being found only in around ten percent of Icelanders (one would suspect that they are considerably more frequent in Scotsmen), which is just as well because they are also major susceptibility factors for skin cancer. Similar mutations are responsible for red hair in all sorts of other mammals, including cats and dogs, and even more rare variants are responsible for albinism in humans. Furthermore, it turns out that at least some Neanderthal Men had a mutation in this gene, leading to speculation that our ancient cave brethren were fair-skinned with red hair and freckles.

If you were to scan for blonde hair color in Americans, I suspect you'd come up with genes for vanity, since so many of our blondes are artificially so. But in Europe, variation in all four of these genes plus another one called *KITLG* combines to allow fairly good prediction of whether a person is blonde or brunette. They're better at excluding the latter than establishing the former.

This type of result has fascinating, and perhaps troubling, implications for forensics. It raises the possibility that in the not too distant future police may begin to profile their suspects on the basis of features such as hair or eye color as implied by the contents of a blood drop. General facial features are likely to be too genetically complex to allow forensic scientists to sketch a face without a witness, but features such as detached earlobes might not be far off either.

We will certainly be looking at racial profiling in a whole new light, since a substantial fraction of the variation in skin pigmentation is attributable to these same genes mentioned previously, working together with a few other genes. Looking at the patterns of variation in the DNA it is pretty clear that lighter skin has been under selection in northern latitudes for thousands of years. The general consensus is that dark skin inhibits the absorption of ultraviolet radiation needed for the manufacture of vitamin D3, but this disadvantage is offset by its protective role against sunburn and skin cancer, now that we're a largely hairless species. More recent speculation has it that the great pigmentation shift actually did not occur coincidentally with human migration into northern Europe, but rather awaited the transition to ways of farming that also shifted our vitamin needs. In Australia, where kids have been encouraged to slip on a shirt, slop on some sunscreen, and slap on a hat to ward off skin cancer, osteoporosis due to vitamin D deficiency is on an alarming rise.

the God gene

Not to belittle their powers, but crystal balls, tea leaves, and palm reading leave a lot to be desired when it comes to accurate prediction of the future. The question before us now is whether personal genotyping will prove to be any more accurate. It is one thing to fork out $20 for a séance on the Las Vegas strip, another to find $1,000 to send off to "23andMe" or "DecodeMe" for a saliva kit that gets your own genome profiled using the same methods that underlie much of the research in this book. I'm toying with the idea myself, even though I know the money could be better spent paying down the mortgage, or even better, providing heifers for Haitian villagers.

To be fair, the objective of these new personal genome ventures is to provide people with an accurate read out of their ancestry. But our susceptibility to disease, along with some portion of our predilections for whom we become, is inscribed in the very same code. As the next few years unfold, companies will come online offering to interpret your personal genomes, providing a glimpse of the future for you and your kids. Will they be short or tall, extroverted, disagreeable, or emotionally unstable, athletic or beautiful? Well, of course they will be beautiful and smart, but you get my gist.

The first person to head in this direction is the first person to have had his entire genome sequenced, none other than Craig Venter, the founder of Celera Genomics. He published a paper describing the millions of genetic variants in his DNA in late 2007, and then incorporated two dozen snippets of this information as boxes within his fascinating autobiography, A *Life Decoded*. If you read these closely, you will get a sense of how impossibly silly this genetic prediction business is for now. For example, he sees some agreement in the sequence of his *GSTM1* and *CYP1A2* genes and his sensitivity to asthma and ability to thrive on caffeine, respectively, but also notes inconsistency between his *PER3* and *DRD4* alleles and his tendency to be a night owl and to embark on risky ventures. Regarding the diseases we've discussed in this book, his genome is a perfectly average mixed blessing: Some hint of Alzheimer's susceptibility here is offset by a protective variant there; his outlook for depression and coronary heart disease are about normal; and diabetes yields a completely unconvincing read out as well.

In each case, Venter wryly remarks that, well, a single gene never tells the whole story, and more research is needed. If disingenuous, at least it is honest, more so to my mind than the approach of the second person to have his genome sequenced, James D. Watson of double helix fame. A rival company with a new technology, 454 Life Sciences, has compiled his famous genome, but certain parts will be masked from public release, ostensibly to protect the privacy of his children. Oh the painful decisions those future celebrities will face as a hungry public clamors for information about what makes them special: Will *People* magazine soon be publishing genome updates on Paris Hilton's addictions and Tom Cruise's fertility?

From memory to conscientiousness, psychologists now agree that the genome has some role to play in human behavior. We can be sure that genome scans for specific behaviors will soon be published, and equally sure that they will identify no more than a handful of genes that explain no more than a few percent of the variability that we experience. Further, we will see a rash of popular articles arguing just why this or that particular variant has been selected by evolution because it is good to be aggressive or altruistic, zany or zealous.

But behavior is more complex even than the diseases we've considered. While I have no doubt that bits of DNA are associated with all these behaviors, I am equally certain that we have little hope of unraveling just what balance of forces shape their prevalence in the human gene pool. Beware then of simplistic explanations for one person's foibles and be skeptical of rationalizations on the basis of some specific ancestral human need.

It would be equally unwise to completely dismiss the idea that attributes as personal as spirituality are somehow embedded in our genomes. Dean Hamer has written the first book on this topic. Provocatively titled *The God Gene: How Faith Is Hardwired into Our Genes*, it lays out research conducted in his own lab that purports to find a gene that influences how religious a person is. Read the last chapter, though: If anything, the wiring is extremely soft, the gene implicated has just a small effect, and it is definitely neither a gene for which religion a person follows, nor even for whether they are religious, actively praying to a public God. Rather, the book describes an allele that correlates with a psychological dimension that has something to do with how spiritual a person is, spirituality being more of a Dalai Lama trait than a Jerry Falwell one.

Here's how the strategy works, pretty much following how geneticists go about finding disease-associated genes, except that some measure of godliness is substituted for disease. You find a large number of unrelated people and ask them a bunch of questions designed to reveal their inner self on a five-point scale. For example, if you answer "I would prefer to be at the office rather than home with my new baby" with the "Strongly Agree" option, you're probably disagreeable, while answering "Mildly Disagree" might knock points off your conscientiousness score. It turns out that a particular set of 30 or so such questions can be used to devise a single score that after the fact has been claimed to measure spirituality. The score also shows a high correlation between parents and their kids, showing that whatever the score is measuring is heritable. Then you just comb through suspect places in the genome for sites that have an A instead of a G, for example, in people with different spiritual values. Lo and behold a gene called *VMAT2* has a variant that is a little bit enriched in your basic Dalai Lama types.

I find it ironic that Hamer used basically the same approach to argue that he had found a gene for homosexuality a few years back as well, notably with equally controversial consequences. And equally nebulous association: Neither of these results has been extensively replicated. And although one or both may be true, it seems likely that just like the genetic variants that influence height, these do no more than gently nudge a person in a certain direction.

There are in fact literally thousands of similar results in the behavioral genetics literature. Perhaps the most famous one is a change in front of one of the dopamine receptor genes, which has been found to be enriched in people who engage in all manner of risky behaviors from jumping out of planes to participating in orgies or exploring drugs. The thing about all these studies is that they are performed out of context of the rest of the genome, and now that we have the tools to study every gene all at once, their effects will be seen to be a tiny part of the complicated whole. They are like the corner pieces of a jigsaw puzzle, convenient anchor points, but alone they don't even tell you what the picture is about.

a few words about IQ

Inevitably, pundits are going to start noticing that some of the variants supposedly associated with human differences are at different frequencies in different racial groups. Nowhere is this more worrying than in relation to inference about human intelligence. In fact, it has already started. *Microcephalin,* for example, is a gene required for brain development in mice, and rare mutations also cause human children to be born with tiny brains. It has been argued that it is one of the most strongly selected genes in humans, with particular alleles being prominent in Caucasians but absent in Africans. Ergo, in the minds of those quick to jump to flimsy conclusions, Africans are said to be genetically inferior intellectually. In fact, when others went and actually looked, they found absolutely no association between the selected allele and IQ.

There is a long history of attempts to establish that African Americans have a genetically half full cupboard, but intriguingly little attempt to similarly argue the other way for Asians. The basic argument is that IQ testing has repeatedly demonstrated significant differences in IQ between ethnic groups, while testing of twins has pretty convincingly shown that there is a strong genetic component to IQ. Ergo, the difference between the races must be genetic—but for a simple error of logic. Just because the variation within a group has one cause does not imply that the difference between groups has the same cause. No one has a problem understanding this when the discussion is about the height of immigrants who move to a developed country: The average height of their children typically jumps a couple of inches. Obviously the genes didn't change, so the difference between the generations must be dietary. So why are many people so quick to attribute racial differences to purely hypothetical genes?

Now consider some of the human traits that are most commonly transmitted across generations: religious observance, allegiance to sporting teams, and—believe it or not—a love of opera. I would imagine that most of us have a hard time believing that genes are the least bit involved in any of these behaviors. Maybe they have something to do with musical appreciation, but clearly these examples of transmission can be explained by cultural and geographic factors that just happen to be correlated with familial relationships. Inheritance, in a broad and colloquial sense, refers to the transfer of something, such as property or assets, from a parent or guardian to a child. Since genes are responsible for the transmission of biological attributes, it has come to be assumed that if there is resemblance between parents and offspring, genes must be involved. But Muslims tend to beget Muslims and Catholics beget Catholics, and yet none of us would rightly conclude that there are genes for Islam and Christianity. The point is that the combination of education and the environment is much more likely than genetics to account for the differences. We need more data of a different sort to establish that genes are responsible.

It is thus quite possible for any trait to be highly heritable, and to differ greatly between two groups, but for this difference to have nothing to do with the genes. Anyone who follows the syllogism above, that genetic variation within a group implies genetic divergence between them, is selling you a bill of goods. So long as the environment can bring about a difference between peoples, it is risky to invoke genetic differences and appropriate to demand an exceptional standard of proof when people's lives are at stake.

Nevertheless, bloggers are starting to notice that there are genetic variants that appear to be associated with intelligence or other behaviors and appear to differ in frequency between races. How should we react to reports that scientists have found some genetic variant in 70 percent of Whites that is associated with IQ score at a probability of 0.000001, and is present in just a minority of Blacks?

First, recognize that significance values alone don't say anything about the magnitude of an effect. Throughout this book we have discussed genetic associations that are significant at probabilities thousands of times more convincing than this, yet only explain a few percent of susceptibility to a disease. These results need to be replicated multiple times in many tens of thousands of people before they become convincing. Typically, when this is done, the estimated magnitude of the effect becomes less and less. For now it seems highly unlikely that any single difference will account for more than an IQ point or two, whatever that means. Miniscule in relation to all the incredible potential bottled up in all our genomes.

Second, ask whether the experiment has been done in the reverse direction. Have they looked for associations with IQ in other races and then asked whether the alleles are at a low frequency in Whites? The trouble is that there is an enormous ascertainment bias in the way genetics is done predominantly on northern Europeans and affluent Americans. You might find dozens of associations that involve alleles that are enriched in these populations, but they say nothing about what variants might be present in the other racial groups and are lacking in Caucasians. The work has to be balanced. As it happens, Africans are on the whole considerably more genetically diverse than any other human group, and this fact alone suggests that a certain amount of genetic potential has been lost in the course of migration around the globe.

Third, put the research in context. Another theme of this book has been that the environment modulates genetic effects, often swamping them. A case in point here is a just-published study from the group that brought us the interaction between stress and a serotonin transporter variant that influences suicidal tendencies. Breast-feeding, it is suggested, improves childhood IQ measurement, particularly in children who can digest particular fatty acids because they have a particular flavor of the *FADS2* gene. Breast-feeding? Colored balls above the crib I can believe, but how many other subtle and not-so-subtle parental behaviors must affect how we develop?

Finally, keep in mind that average values are meaningless when it comes to appreciating the worth of individuals. Even if there were a genetically based difference of a few points in IQ between ethnic, cultural, or geographic groups of humans, the brightest tier of the group with the lower average would still score higher than almost everyone in the other one, and almost half of the first group would be genetically as well if not better endowed than every other member of the second. How many of us know our own IQ within 10 points, let alone those of our friends and colleagues?

Genes play their part, but let's attend to the things we can control. It is perverse to think that a society would focus on the relatively small component that may be genetic, rather than doing everything in its power to maximize the potential of every individual by doing something about the much larger component that it is within our power to address, namely education and public health.

on being human

What is it that makes us humans human? What are the genetic changes that have made us a species that looks up at the stars and sees future colonies instead of pagan gods, that spends its free time picking out friends on Facebook instead of picking at gnats on friends, and that thrives in the jungles of Manhattan as ably as it roams the African savannah?

Surely we should have some answers by now. If we compare the blueprints of a Porsche Boxster and a Honda Civic side by side, it is immediately apparent how they come to differ. Similarly, you might think that if we align the genomes of a human and a chimpanzee, it ought to be straightforward to stitch together the evolutionary path that leads to *Homo sapiens sapiens*.

Superficially, it is. Our capacity for smell is different from other primates, since our repertoire of olfactory receptors, the molecules that sense perfumes and rotten eggs alike, looks more unlike that of our nearest relatives than do most genes. Similarly, the sequences of our immune system receptors are in general quite diverged, much as the bumper bars on a Porsche and a Honda are barely recognizable variations on a theme. Neither of these changes is remarkable, and both are entirely predictable given a moment's thought.

There are actually a half dozen subtly different ways to compare primate genomes, but they all tell pretty much the same story. Namely that hundreds if not thousands of genes are suspects of interest in the making of modern humans. Some have molecular roles in causing neurons to divide (so that the human brain got bigger), some look like they digest toxic compounds (so our livers can tolerate wild herbs in the new world), some are involved in deposition of calcium in the bones, in hearing, in pigmentation, and...you name it.

Intriguingly, many are also linked to human diseases. Among those 25 growth genes that give or take a millimeter or two of height, are several also implicated in cancer and others that make a contribution to osteoporosis. When a scan was just completed for nicotine dependence, it found a gene that is also the major genetic risk factor for lung cancer. Partly because a brain that desires more cigarettes exposes it's body to more carcinogens, but also, it seems, because the nicotine receptor it encodes is also active in the lungs. The journey from normality to disease susceptibility takes many twists and turns, but results such as these show just how intimate the link between human evolution and susceptibility to disease is.

What is striking to me, though, is that there has been no eureka moment when as biologists we have been able to look at this extraordinary new data and realize "That's it! That is the genetic switch for Homo sapienism. If that mutation over on chromosome 14 hadn't occurred, we'd still all be back in Olduvai wondering whether we can make a better life for our families by swinging through trees or wandering the plains. Or without this other mutation, no one would be composing concertos or smashing baseballs out of Yankee Stadium." No, just like every other species on the planet, we humans are a product of millions of little tinkerings. The secret to understanding our humanity is not so much in the individual genes as in the way those genes interact with one another as a genome.

the adolescent genome revisited

As change is a hallmark of adolescence, so adolescence is a defining characteristic of the human genome. Change is ever present in the history of our species. We see it in the genetic upheavals that must have accompanied the origins of the genus a million years ago, and of the species 150,000 years ago. We also see it in the myriad ongoing genetic turnovers that have occurred as we have populated every nook and cranny of the planet. We see change in the environments and cultures that humans have occupied in their transitions from nomad to pastoralist to city-dweller, these too occurring over timescales that range from tens of thousands to just tens of years. And we see change in the distribution of disease prevalence, particularly with the rise of the complex diseases that will claim the great majority of our lives.

This book has been an argument that these three modes of change are deeply connected. Specifically, the combination of genetic and environmental change has given rise to modern disease susceptibility. It is a more subtle formulation than the trite assessment that a gene makes you sick as a side effect of some benefit it also confers. To be sure, sometimes this is the case, but a more critical look tells us that just like a teenager, the genome is trying mightily to come to grips with its growing pains.

At this point, I need to confess that this formulation, that "genetic + environmental change = increased disease susceptibility" falls well short of a syllogism. There is no logical reason why the right-hand side of the equation should follow from the left-hand side, and patently in many instances it does not. Genius exhibited by musicians and scientists, novelists and entrepreneurs is equally a product of the combination of the genetic evolution of the brain with centuries of cultural advance. The argument, then, is not so much that A + B = C, as that C is dependent on and explained by A and B. Without the twin sources of change, we should not expect to see so much human suffering at the whim of our genomes.

The full argument is more sophisticated, but I have refrained from developing it but for occasional asides about canalization. In a nutshell, this is the notion that over millions of years, species evolve not just toward their genetic optima, but also to ensure that they are well buffered and robust, resistant to all kinds of perturbations. When change comes on a massive scale, it makes everything more variable than it would be under normal circumstances. And regarding health, that heightened variability is seen as susceptibility to each of the complex diseases for 10 or 20 percent of the population, instead of just a small minority.

This reasoning is familiar enough to evolutionary biologists, but as yet holds little currency with biomedical geneticists—not for want of truth, but rather because clinicians are more concerned with the practical problem of finding the genes that contribute to disease. Only in the last couple of years have they been granted the tools with which to conduct their searches in a systematic manner. The first flushes of success have generated extraordinary excitement in the research community, but they come with a tinge of disappointment as well, because it is immediately apparent that simple answers will not be forthcoming.

Each of the half dozen or so genes that we have discussed in each of the chapters of this book explain only a small percentage of the reason why some people get diabetes, depression, and so forth, while others do not. Over the next few years, hundreds of millions of dollars will be spent and hundreds of thousands of people will have their genomes scanned, as a result of which the number of genes implicated in each disease will likely rise to two dozen or more. Yet even then there is little reason to believe we will be a lot closer to explaining why specifically your relatives are more likely to suffer the same diseases as you.

For those who might be hoping for some sort of test to be taken at birth that will tell a person just which of the common diseases to look out for, don't hold your breath. For the foreseeable future, you're probably going to be just as well off making common-sense inferences from the health of your brothers, sisters, parents, and Great-Aunt Bessies. Ultimately, knowing the genes should help the drug companies develop more effective therapies, and the nutritionists and lifestyle counselors to promote better ways of living. Whether their recommendations will be personalized based on knowledge of your genome is still debatable.

The genetics will tell us, though, why our genes make us sick. Why hundreds of places in the genome influence every disease, and how the environment works with them. Genetics will tell us that some genetic variants are there because they used to be beneficial at an earlier stage of human evolution, but now are bad for us. It will tell us that some have a yin and yang good and bad side to them in today's world. More often it will be apparent that variants are there because they are an unavoidable part of the way life is, their effects suppressed as much as possible, but never perfectly so. And in some cases we will see that risky alleles carried over from earlier times have not yet been replaced by better ones that truly protect our health.

If we come back in a few million years, perhaps we will find that our adolescent genome has evolved to a more mature equilibrium that offers greater protection from disease. In the meantime we must make do with a system that gives way more than it takes. It is after all also responsible for the extravagant and bountiful diversity of the human condition.

Notes

Chapter 1

genetic imperfection

It should not be necessary for readers new to genetics to have an understanding of how genes work, but a good place to start would be Larry Gonick and Mark Wheelis's *The Cartoon Guide to Genetics* (Harper Collins, 1991).

unselfish genes

An interesting online discussion of "gay genes" and how they have been portrayed in mainstream media can be found at www.leaderu.com/jhs/satinover.html.

Of all the excellent books written by Harvard geneticist Richard Lewontin on population genetics, I recommend *Human Diversity* (Scientific American Library, 1991—a general discussion of human variation) and *It Ain't Necessarily So* (Granta Books, 2001—a critique of the Human Genome Project).

The title of this section is a play on Richard Dawkins's popular book *The Selfish Gene* (Oxford University Press, 1978), which paints a very deterministic picture of genes and evolution.

how genes work and why they come in different flavors

For a technical description of *staufen*'s role in memory, see the article by Dubnau, J., A. S. Chiang, L. Grady, J. Barditch, S. Gossweiler, J. McNeil, P. Smith, F. Buldoc, R. Scott, U. Certa, C. Broger, and T. Tully (2003) *Current Biology* 13: 286-296 "The staufen/pumilio pathway is involved in Drosophila long-term memory," and the earlier review by Roegiers, F. and Y-N. Jan (2000) *Trends in Cell Biology* 10: 220-224 "Staufen: a common component of mRNA transport in oocytes and neurons?"

Most papers on mutation-selection-drift balance are highly mathematical and technical, so not for the faint-of-heart. An oft-cited one is Barton, N. H. and M. Turelli (1989) *Annual Review of Genetics* 23: 337-370 "Evolutionary quantitative genetics: how little do we know?" while a more recent contribution is by Zhang, X. S., J. Wang, and W. G. Hill (2002) *Genetics* 161: 419-433 "Pleiotropic model of maintenance of quantitative genetic variation at mutation-selection balance."

There is vast literature on evolutionary psychology and evolutionary medicine, championed by Randolph Nesse in a popular book, *Why We Get Sick: The New Science of Darwinian Medicine* (Knopf, 1994), as well as in the scientific literature:

Nesse, R. M., S. C. Stearns, and G. S. Omenn (2006) *Science* 311: 1071 "Medicine needs evolution." In this book, I espouse quite a different view of the role of genetic evolution in disease.

three reasons why genes might make us sick

Genetic modifiers of cystic fibrosis are discussed in Knowles, M. R. (2006) *Current Opinion in Pulmonary Medicine* 12: 416-421 "Gene modifiers of lung disease," while variation in the *CFTR* gene is described in Rowntree, R. K. and A. Harris (2003) *Annals of Human Genetics* 67: 471-485 "The phenotypic consequences of CFTR mutations."

For an early review of *Dystrophin* see Worton, R. G. and M. W. Thompson (1988) *Annual Review of Genetics* 22: 601-629 "Genetics of Duchenne muscular dystrophy."

Spinocerebellar ataxias and triplet expansion diseases are described in a provocatively titled article by Petronis, A. and J. L. Kennedy (1995) *American Journal of Psychiatry* 152: 164-172 "Unstable genes—unstable mind?"

The difficulties involved in dissecting the genetics of schizophrenia are clearly laid out in Sanders, A. R. et al. (2008) *American Journal of Psychiatry* 165: 497-506 "No significant association of 14 candidate genes with schizophrenia in a large European ancestry sample: Implications for psychiatric genetics."

A standard textbook of quantitative genetics that includes the theory behind threshold liability is Michael Lynch and Bruce Walsh's *Genetics and Analysis of Quantitative Traits* (Sinaeur, 1998).

For a technical review of canalization see Gibson, G. and G. Wagner (2000) *Bioessays* 22: 372-380 "Canalization in evolutionary genetics: a stabilizing theory?" The original observation in mice was described by Dunn, R. B. and A. S. Fraser (1958) *Nature* 181: 1018-1019 "Selection for an invariant character—'vibrissae number'— in the house mouse," and the idea was first popularized by C. H. Waddington in *The Strategy of the Genes* (McMillan, 1957).

One of the best descriptions of modern human evolution is by Luigi Luca and Francesco Cavalli-Sforza, *The Great Human Diasporas: The History of Diversity and Evolution* (Addison Wesley, 1996).

the human genome project

Amazon.com lists nearly 7,000 entries under the heading of "human genome project." Two good starting points are Daniel Kevles and Leroy Hood's *The Code of Codes: Scientific and Social Issues in the Human Genome Project* (Harvard University Press, 1993) and Michael Palladino's *Understanding the Human Genome Project*, 2nd edition (Benjamin Cummings, 2005). Another very readable and interesting summary of the project is Matt Ridley's *Genome: An Autobiography of the Species in 23 Chapters* (Harper Perennial, 2000).

The two papers describing the draft sequence of the human genome were published simultaneously in February 2001. J. C. Venter et al. (2001) *Science* 291: 1304-1351 "The sequence of the human genome," and E. S. Lander et al. (2001) *Nature* 413: 860-921 "Initial sequencing and analysis of the human genome." Both journals provide public Web sites for perusal of the papers and commentaries.

Scientists use three major Web sites to access all the genome project data: the US-NIH site at www.ncbi.nlm.nih.gov, the European molecular biology organization site at www.ensembl.org, and the University of California at Santa Cruz browser at http://genome.ucsc.edu.

The SNP at position 102,221,163 on chromosome 11, also known as rs3025058, is actually in front of a gene called *MMP3* and is known to promote heart disease. See Beyzade, S., S. Zhang, Y-k. Wong, I. Day, P. Eriksson, and S. Ye (2003) *Journal of the American College of Cardiology* 41: 2130-2137 "Influences of *matrix metalloproteinase-3* gene variation on extent of coronary atherosclerosis and risk of myocardial infarction," and Rockman, M. V., M. Hahn, N. Soranzo, D. Loisel, D. B. Goldstein, and G. A. Wray (2004) *Current Biology* 14: 1531-1539 "Positive selection on *MMP3* regulation has shaped heart disease risk."

genomewide association

The study of human genetic variation using the GWA approach garnered *Science* journal's "Breakthrough of the Year" accolade for 2007. While dozens of papers were published, the one that marked a turning point was: Wellcome Trust Case Control Consortium (2007) *Nature* 447: 661-678 "Genome-wide association study of 14,000 cases of seven common diseases and 3,000 shared controls."

Chapter 2

cancer of the breast

Wikipedia is a useful source of information on all of the diseases discussed in the next several chapters. Lists of celebrity survivors and patients can be found in many places on the Internet, such as http://en.wikipedia.org/wiki/List_of_notable_breast _cancer_patients_according_to_survival_status. The full breast cancer site is at http://en.wikipedia.org/wiki/Breast_cancer.

In addition to her *Uplift: Secrets from the Sisterhood of Breast Cancer Survivors* (Simon and Schuster, 2003), Barbara Delinsky has a Web site for survivors at www.barbaradelinsky.com/delinsky-uplift.htm.

The National Cancer Institute's Web site provides updated statistics on the incidence of breast cancer and information about treatment and prognosis at www.nci.nih.gov/cancertopics/types/breast.

broken genes, broken lives

Some recent studies of the influences of socioeconomic status and race on cancer incidence include Chu, K. C., B. A. Miller, and S. A. Springfield (2007) *Journal of the National Medical Association* 99: 1092-1104 "Measures of racial/ethnic health disparities in cancer mortality rates and the influence of socioeconomic status," Albano, J. D., E. Ward, A. Jemal, R. Anderson, V. Cokkinides, T. Murray, J. Henley, J. Liff, and M. J. Thun (2007) *Journal of the National Cancer Institute* 99: 1384-1394 "Cancer mortality in the United States by education level and race," and Baquet, C. R. and P. Commiskey (2000) *Cancer* 88 (Suppl.): 1256-1264 "Socioeconomic factors and breast carcinoma in multicultural women."

epidemiology and relative risk

For more on Janet Lane Claypon see an article by medical historian Warren Winkelstein (2004) *American Journal of Epidemiology* 160: 97-101 "Vignettes of the history of epidemiology: Three firsts by Janet Elizabeth Lane-Claypon" as well as the Wikipedia site at http://en.wikipedia.org/wiki/Janet_Lane-Claypon.

A comprehensive review of global trends in breast cancer rates can be found in Althuis, M. D., J. Dozier, W. Anderson, S. Devesa, and L.A. Brinton (2005) *International Journal of Epidemiology* 34: 405-412 "Global trends in breast cancer incidence and mortality 1973-1997." The Center for Disease Control provides a free online file containing cancer statistics for the United States at http://apps.nccd.cdc.gov/uscs (it is over 8Mb). A site that conveniently allows you to visualize customizable color maps of incidence is at www3.cancer.gov/atlasplus/index.html.

The role of estrogen in breast cancer is reviewed in Clemons, M. and P. Goss (2004) *New England Journal of Medicine* 344: 276-285 "Estrogen and the risk of breast cancer." Many studies have examined the effect of hormone replacement therapy, including one of more than a million women, and a lively discussion of the issues can be found in *The Lancet* (2003) volume 362, issue 9392. Reproductive factors are discussed in N. Andrieu et al. (1998) *British Journal of Cancer* 77: 1525-1536 "The effects of interaction between familial and reproductive factors on breast cancer risk: a combined analysis of seven case-control studies," and in Becher, H., S. Schmidt, and J. Chang-Claude (2003) *International Journal of Epidemiology* 32: 38-48 "Reproductive factors and familial predisposition for breast cancer by age 50 years. A case-control-family study for assessing main effects and possible gene-environment interaction."

brakes, accelerators, and mechanics

An excellent introduction to cancer biology for the general public is Robert A. Weinberg's *One Renegade Cell* (Basic Books, 1999).

There are thousands of reviews of the three different classes of cancer-related genes. Here are some classics from the *Annual Reviews* series: Levine, A. J. (1993) *Annual Review of Biochemistry* 62: 623-651 "The tumor suppressor genes," Riley, D. J., E. Lee, and W. H. Lee (1994) *Annual Review of Cell Biology* 10: 1-29 "The

Retinoblastoma protein: more than a tumor suppressor," Knudson, A. G. (2000) *Annual Review of Genetics* 34: 1-19 "Chasing the cancer demon," Bishop, J. M. (1983) *Annual Review of Biochemistry* 52: 301-354 "Cellular oncogenes and retroviruses," Varmus, H. E. (1984) *Annual Review of Genetics* 18: 553-612 "The molecular genetics of cellular oncogenes," and McKinnon, P. J. and K. W. Caldecott (2007) *Annual Review of Genomics and Human Genetics* 8: 37-55 "DNA strand break repair and genetic disease." The genetics of colon cancer is reviewed in de la Chapelle, A. (2004) *Nature Reviews Cancer* 4: 769-780 "Genetic predisposition to colorectal cancer."

Arthur Knudson's seminal paper developing the two-hit model for retinoblastoma was published in 1971 as *Proceedings of the National Academy of Science (USA)* 68: 820-823 "Mutation and cancer: statistical study of retinoblastoma." He wrote a personal account of his life in science a couple of years ago: Knudson, A. G. (2005) *Annual Review of Genomics and Human Genetics* 6: 1-14 "A personal sixty-year tour of genetics and medicine."

Richard Dawkins's second famous book about evolution is *The Blind Watchmaker* (Penguin, 1990, published in the United States by Norton).

familial breast cancer

The contributions of BRCA1 and BRCA2 were estimated by King, M. C., J. Marks, J. Mandell, and the New York Breast Cancer Study Group (2003) *Science* 302: 643-646 "Breast and ovarian cancer risks due to inherited mutations in BRCA1 and BRCA2" and have been more recently reviewed by Fackenthal, J. D. and O. I. Olopade (2007) *Nature Reviews Cancer* 7: 937-948 "Breast cancer risk associated with BRCA1 and BRCA2 in diverse populations."

A fact sheet about genetic testing for BRCA genes is provided by the National Cancer Institute at www.nci.nih.gov/cancertopics/factsheet/Risk/BRCA.

The role of CHEK2 in hereditary breast cancer is discussed in Bogdanova, N., S. Feshchenko, C. Cybulski, and T. Dörk (2007) *Journal of Clinical Oncology* 25: 26e "CHEK2 mutation and hereditary breast cancer." Also see Walsh, T. and M. C. King (2007) *Cancer Cell* 11: 103-105 "Ten genes for inherited breast cancer."

The distribution of BRCA1 mutations and high prevalence in Ashkenazi is reported in John, E. M., A. Miron, G. Gong, A. Phipps, A. Felberg, F. Li, D. West, and A. S. Whittemore (2007) *Journal of the American Medical Association* 298: 2869-2876 "Prevalence of pathogenic BRCA1 mutation carriers in 5 US racial/ethnic groups."

growth factors and the risk to populations

There are hundreds of studies of association of one or several genes with aspects of breast cancer. The following take a more comprehensive genomewide look and are discussed in the text:

A. Cox et al. (2007) *Nature Genetics* 39: 352-358 "A common coding variant in CASP8 is associated with breast cancer risk."

Pharoah, P. D., J. Tyrer, A. M. Dunning, D. F. Easton, B. A. Ponder, and the SEARCH Investigators (2007) *PLoS Genetics* 3: e42 "Association between common variation in 120 candidate genes and breast cancer risk." See also Breast Cancer Association Consortium (2006) *Journal of the National Cancer Institute* 98: 1382-1396 "Commonly studied single-nucleotide polymorphisms and breast cancer: results from the Breast Cancer Association Consortium."

D. F. Easton et al. (2007) *Nature* 447: 1087-1093 "Genome-wide association study identifies novel breast cancer susceptibility loci," D. J. Hunter et al. (2007) *Nature Genetics* 39: 870-874 "A genome-wide association study identifies alleles in *FGFR2* associated with risk of sporadic postmenopausal breast cancer," and S. N. Stacey et al. (2007) *Nature Genetics* 39: 865-869 "Common variants on chromosomes 2q35 and 16q12 confer susceptibility to estrogen receptor-positive breast cancer."

pharmacogenetics and breast cancer

You can download a copy of Senator Obama's bill at www.personalizedmedicinecoalition.org/sciencepolicy/public-policy_senator.php.

The use of *TPMT* genotyping to accompany drug administration is discussed in Maitland, M. L., K. Vasisht, and M. J. Ratain (2006) *Trends in Pharmacological Science* 27: 432-437 "TPMT, UGT1A1 and DPYD: genotyping to ensure safer cancer therapy?"

A very recent review of emerging cancer therapies is Doyle, D. M. and K. D. Miller (2008) *Breast Cancer* 15: 49-56 "Development of new targeted therapies for breast cancer."

For a brief history of tamoxifen, see http://en.wikipedia.org/wiki/Tamoxifen. New insight into the mechanism of tamoxifen resistance can be read about in Massarweh, S., C. Osborne, C. Creighton, L. Qin, A. Tsimelzon, S. Huang, H. Weiss, M. Rimawi, and R. Schiff (2008) *Cancer Research* 68: 826-833 "Tamoxifen resistance in breast tumors is driven by growth factor receptor signaling with repression of classic estrogen receptor genomic function."

Testing of Her2 status remains controversial but is recommended by Carlson, R. W. and the NCCN HER2 Testing in Breast Cancer Task Force (2006) *Journal of the National Comprehensive Cancer Network* 4 (Suppl 3): S1-22 "HER2 testing in breast cancer: NCCN Task Force report and recommendations." Genentech's drug, Herceptin (trastuzumab) is described at www.gene.com/gene/products/information/oncology/herceptin.

The first study using microarrays to perform gene expression profiling of blood cancers and suggesting that they may be predictive of long-term prognosis was A. A. Alizadeh et al. (2000) *Nature* 403: 503-511 "Distinct types of diffuse large B-cell lymphoma identified by gene expression profiling." This approach was demonstrated for breast cancer by L.J. van 't Veer, *et al.* (2002) *Nature* 415: 530-536 "Gene expression profiling predicts clinical outcome of breast cancer."

Similar work has led to the development of a new test, Oncotype-DX, that is undergoing clinical evaluation. A report on this is L. A. Habel et al. (2006) *Breast Cancer Research* 8: R25 "A population-based study of tumor gene expression and risk of breast cancer death among lymph node-negative patients." See the Web site at www.genomichealth.com/oncotype/default.aspx.

why do genes give us cancer?

Darwinian medicine arguments for cancer are advanced by Mel Greaves in (2006) *Nature Reviews Cancer* 7: 213-221 "Darwinian medicine: a case for cancer," and (2002) *Lancet Oncology* 3: 244-250 "Cancer causation: the Darwinian downside of past success?" The same author has written a popular book on the topic, *Cancer: The Evolutionary Legacy* (Oxford University Press, 2001).

Chapter 3

jackie and ella

A list of celebrity diabetics can be found on the Islets of Hope Web site at www.isletsofhope.com/family/famous_diabetics_1.html.

For more information on the life of Jackie Robinson, see his official Web site at www.jackierobinson.com/ or visit his Foundation at www.jackierobinson.org.

Ella Fitzgerald's official Web site is www.ellafitzgerald.com/ where the quote attributable to Jimmy Rowles can be found. The link to her charitable foundation is www.ellafitzgeraldfoundation.org/.

the pathology of diabetes

Much basic information on the disease can be accessed through Wikipedia (http://en.wikipedia.org/wiki/Diabetes_mellitus), the World Health Organization (www.who.int/diabetes/en/), and the Centers for Disease Control (www.cdc.gov/diabetes/). The NIH's National Institute of Diabetes and Digestive and Kidney Diseases also has an information clearinghouse at http://diabetes.niddk.nih.gov/.

Statistics on incidence of diabetes can be found at www.cdc.gov/diabetes/statistics/incidence/.

type 1 diabetes

The genetics of diabetes is reviewed in Florez, J. C., J. Hirschhorn, and D. Altshuler (2003) *Annual Review of Genomics and Human Genetics* 4: 257-291 "The inherited basis of Diabetes mellitus: Implications for the genetic analysis of complex traits," and of T1D specifically in Pociot, F. and M. F. McDermott (2002) *Genes and Immunity* 3: 235-249 "Genetics of type 1 diabetes mellitus," as well as Kim, M. S. and C. Polychronakos (2005) *Hormone Research* 64: 180-188 "Immunogenetics of type 1 diabetes." A large linkage scan is summarized by P. Concannon and the Type

1 Diabetes Genetics Consortium (2005) *Diabetes* 54: 2995-3001 "Type 1 diabetes: evidence for susceptibility loci from four genome-wide linkage scans in 1,435 multiplex families."

Genomewide association studies for T1D are described by J. Todd et al. (2007) *Nature Genetics* 39: 857-864 "Robust associations of four new chromosome regions from genome-wide analyses of type 1 diabetes," and H. Hakonarson et al. (2007) *Nature* 448: 591-594 "A genome-wide association study identifies KIAA0350 as a type 1 diabetes gene." The Wellcome Trust Case Control Consortium (2007) *Nature* 447: 661-678 "Genome-wide association study of 14,000 cases of seven common diseases and 3,000 shared controls" also includes both T1D and T2D.

For updates on the role of the HLA complex in promoting T1D and other autoimmune diseases, see Lie, B. A. and E. Thorsby (2005) *Current Opinion in Immunology* 17: 526-531 "Several genes in the extended human MHC contribute to predisposition to autoimmune diseases," and Larsen. C. E. and C. A. Alper (2004) *Current Opinion in Immunology* 16: 660-667 "The genetics of HLA-associated disease."

There is vast literature on the relationship between breast-feeding, baby formula, and both types of diabetes. Some of the studies in support of an association are an Australian study by H. Malcova et al. (2006) *European Journal of Pediatrics* 165: 114-119 "Absence of breast-feeding is associated with the risk of type 1 diabetes: a case-control study in a population with rapidly increasing incidence," a Finnish one by S. M. Virtanen et al. (1992) *Diabetic Medicine* 9: 815-819 "Feeding in infancy and the risk of type 1 diabetes mellitus in Finnish children," and an English one by P. A. McKinney et al. (1999) *Diabetes Care* 22: 928-932 "Perinatal and neonatal determinants of childhood type 1 diabetes. A case-control study in Yorkshire, U.K." However, for a much more skeptical meta-analysis see Norris, J. M. and F. W. Scott (1996) *Epidemiology* 7: 87-92 "A meta-analysis of infant diet and insulin-dependent diabetes mellitus: do biases play a role?" For a comprehensive discussion of the mechanisms by which nutrition and genes may influence T1D, see Karges, W. J. P., J. Ilonen, B. H. Robinson, and H.M. Dosch (1995) *Molecular Aspects of Medicine* 16: 79-213 "Self and non-self antigen in diabetic autoimmunity: molecules and mechanisms."

The Insulin gene VNTR polymorphism, as well as *PTPN22* and *CTLA4*, are discussed in Anjos, S. and C. Polychronakos (2004) *Molecular Genetics and Metabolism* 81: 187-195 "Mechanisms of genetic susceptibility to type I diabetes: beyond HLA," and Jahromi, M. M. and G. S. Eisenbarth (2006) *Annals of the New York Academy of Science* 1079: 289-299 "Genetic determinants of type 1 diabetes across populations."

SUMO4 is reviewed in Wang, C. Y. and J. X. She (2008) *Diabetes/metabolism Research and Reviews* 24: 93-102 "SUMO4 and its role in type 1 diabetes pathogenesis."

an epidemic genetic disease

Michael Pollan's excellent book is *The Omnivore's Dilemma: A Natural History of Four Meals* (Penguin, 2007).

The politics of corn subsidies has just become yet more interesting with the growth of interest in bioethanol. See these recent articles in the *Washington Post* and *New York Times*: www.washingtonpost.com/wp-dyn/content/article/2007/09/27/AR2007092702054_pf.html and www.nytimes.com/2005/11/09/business/09harvest.html.

genetics of obesity

For a review of leptin in humans, see E. Jéquier (2002) *Annals of the New York Academy of Science* 967: 379-388 "Leptin signaling, adiposity, and energy balance." More gut hormones are reviewed in Murphy, K. G. and S. R. Bloom (2006) *Nature* 444: 854-859 "Gut hormones and the regulation of energy homeostasis."

A guide to weight loss drugs that have been approved for use in the United States can be found at www.healthnetwork.com.au/weight-loss/drugs.asp. This is one area where there is an enormous amount of disinformation on the Web, but sites such as WebMD may help: www.webmd.com/diet/guide/weight-loss-prescription-weight-loss-medicine.

The various hormonal mechanisms of appetite regulation are described by E. T. Rolls (2007) *Obesity Reviews* 8 (Suppl. 1): 67-72 "Understanding the mechanisms of food intake and obesity." For a broader view of the reasons for the onset of T2, see Kahn, S. E., R. L. Hull and K. M. Utzschneider (2006) *Nature* 444: 840-846 "Mechanisms linking obesity to insulin resistance and type 2 diabetes."

The Obesity Gene Map database of all genes implicated in the condition is available online at http://obesitygene.pbrc.edu/ and in print through T. Rankinen et al. (2006) *Obesity* 14: 529-644 "The human obesity gene map: the 2005 update."

The association of FTO with obesity was first reported by L. J. Scott et al. (2007) *Science* 316: 1341-1345 "A genome-wide association study of type 2 diabetes in Finns detects multiple susceptibility variants," C. Dina et al. (2007) *Nature Genetics* 39: 724-726 "Variation in FTO contributes to childhood obesity and severe adult obesity," and A. Scuteri et al. (2007) *PLoS Genetics* 3: e115 "Genome-Wide Association Scan Shows Genetic Variants in the FTO Gene Are Associated with Obesity-Related Traits." The first insights into the molecular function of the gene were reported in: T. Gerken et al. (2007) *Science* 318: 1469-1472 "The obesity-associated FTO gene encodes a 2-oxoglutarate-dependent nucleic acid demethylase."

The *INSIG2* association was first reported by A. Herbert et al. (2006) *Science* 312: 279-283 "A common genetic variant is associated with adult and childhood obesity." More data can be found in H. N. Lyon et al. (2007) *PLoS Genetics* 3: e61 "The association of a SNP upstream of *INSIG2* with Body Mass Index is reproduced in several but not all cohorts." A critical review of *ENPP1* that references several studies is H. N. Lyon *et al.* (2006) *Diabetes* 55: 3180-3184 "Common variants in the ENPP1 gene are not reproducibly associated with diabetes or obesity." Melanocortin receptors are reviewed by R. D. Cone (2006) *Endocrine Reviews* 27: 736-749. Studies on the physiological functions of the melanocortin system and the association with obesity can be found in I. M. Heid et al. (2005) *Journal of Medical Genetics* 42: 21-26 "Association of the 103I *MC4R* allele with decreased body mass in 7937 participants of two population based surveys."

type 2 diabetes

TCF7L2 turned up in several genomewide scans: V. Steinsthorsdottir et al. (2007) *Nature Genetics* 39: 770-775 "A variant in CDKAL1 influences insulin response and risk of type 2 diabetes," L. J. Scott *et al.* (2007) *Science* 316: 1341-1345 "A genome-wide association study of type 2 diabetes in Finns detects multiple susceptibility variants," J. T. Salonen et al. (2007) *American Journal of Human Genetics* 81: 338-345 "Type 2 diabetes whole-genome association study in four populations: the DiaGen consortium." It has been further studied in dozens of places around the world, while A. Helgason et al. (2007) *Nature Genetics* 39: 218-225 "Refining the impact of *TCF7L2* gene variants on type 2 diabetes and adaptive evolution" describe the recent history of the gene.

The initial paper introducing the thrifty genes hypothesis was J. V. Neel (1962) *American Journal of Human Genetics* 14: 353-362 "Diabetes mellitus: a 'thrifty' genotype rendered detrimental by 'progress'?" One of the strongest critiques of it is J. R. Speakman (2006) *Diabetes and Vascular Disease Research* 3: 7-11 "Thrifty genes for obesity and the metabolic syndrome—time to call off the search?" Jared Diamond discusses the notion in *Nature* 423: 599-602 "The double puzzle of diabetes." His two books mentioned in the text are *Guns, Germs, and Steel: The Fates of Human Societies* (Norton, 1997) and *Collapse: How Societies Choose to Fail or Succeed* (Viking, 2005).

Recent work on the evolution of lactose tolerance is described in S. A. Tishkoff et al. (2006) *Nature Genetics* 39: 31-40 "Convergent adaptation of human lactase persistence in Africa and Europe." See also my commentary: Gibson, G. (2007) *Current Biology* 17: R295-R296 "Human evolution: thrifty genes and the dairy queen."

A strong argument that much of the susceptibility to human disease can be attributed to ancient alleles shared with other primates can be found in Di Rienzo, A. and R. R. Hudson (2005) *Trends in Genetics* 21: 596-601 "An evolutionary framework for common diseases: the ancestral-susceptibility model."

disequilibrium and metabolic syndrome

Another recent book also introduces the idea that disequilibrium is responsible for diabetes: Peter Gluckman and Mark Hanson's *Mismatch: Why Our World No Longer Fits Our Bodies* (Oxford University Press, 2006). It enhances the thrifty phenotype hypothesis first introduced by Hales, C. N. and D. J. Barker (2001) *British Medical Bulletin* 60: 5-20 "The thrifty phenotype hypothesis," a notion that is gaining more traction as it becomes apparent that maternal health during pregnancy impacts T2D susceptibility. A more current update including the concept of epigenetic reprogramming can be found in de Moura, E. G. and M. C. Passos

(2005) *Bioscience Reports* 25: 251-269 "Neonatal programming of body weight regulation and energetic metabolism."

Metabolic syndrome is a somewhat controversial concept, but it is estimated that well more than a third of all Westerners suffer from the joint predisposition to diabetes and coronary heart disease due to metabolic problems. See J. B. Meigs (2002) *American Journal of Managed Care* 8(Suppl): S283-S292 "Epidemiology of the metabolic syndrome, 2002," and Batsis, J. A., R. E. Nieto-Martinez, and F. Lopez-Jimenez (2007) *Clinical Pharmacology and Therapeutics* 82: 509-524 "Metabolic syndrome: from global epidemiology to individualized medicine." The connection with obesity is explored in Després, J.-P. and I. Lemieux (2006) *Nature* 444: 881-887 "Abdominal obesity and metabolic syndrome."

Chapter 4

athletic asthmatics

For information about asthma education, see www.asthmaactionamerica.com/. The Asthma All-Stars and other famous people with asthma are introduced at www.healthsmart.org/ibreathe/2_0_asthma/2_2_6_famous_people.htm.

Jackie Joyner-Kersey has a Foundation at http://jackiejoyner-kerseefoundation.org; Jerome Bettis's "The Bus Stops Here" Foundation is at www.thebus36.com/foundation/foundation.htm.

The CDC's statistics on asthma can be found at www.cdc.gov/asthma/ while the World Health Organization site is www.who.int/topics/asthma/en/.

inflammation and respiration

The first review of canine atopic dermatitis that I am aware of is by R. E. Halliwell (1971) *The Veterinary Record* 89:209-214 "Atopic disease in the dog," as referenced in Hillier, A. and C. E. Griffin (2001) *Veterinary Immunology and Immunopathology* 81: 147-151 "The ACVD task force on canine atopic dermatitis (I): incidence and prevalence."

For a brief description of the common asthma drugs, see www.healthcentral.com/asthma/find-drug.html.

A couple of recent reviews of the genetics and immunology behind asthma are Zhang, J., P. D. Pare, and A. J. Sandford (2007) *Respiratory Research* 9:4 "Recent advances in asthma genetics," and Yamashita, M., A. Onodera, and T. Nakayama (2007) *Critical Reviews in Immunology* 27: 539-546 "Immune mechanisms of allergic airway disease: regulation by transcription factors."

the hygiene hypothesis

David Strachan's two papers on the hygiene hypothesis are D. P. Strachan (1989) *British Medical Journal* 299: 1259-1260 "Hay fever, hygiene, and household size," and D.P. Strachan (2000) *Thorax* 55 (Suppl 1): S2-10 "Family size, infection and atopy: the first decade of the 'hygiene hypothesis.'" One of several recent reviews is by E. von Mutius (2007) *Immunobiology* 212: 433-439 "Allergies, infections and the hygiene hypothesis—the epidemiological evidence." Additionally, hundreds of papers examine specific environmental influences on asthma susceptibility.

A fascinating read about polio is David M. Oshinsky's *Polio: An American Story* (Oxford University Press, 2006). For a history of the March of Dimes, try David W. Rose's *March of Dimes: Images of America* (Arcadia Books, 2003).

asthma epidemiology

Prevalence by ethnicity and age in the United States in adults is around 11 percent for Caucasians and Africans, but less than 8 percent for Hispanics, and more than 20 percent in Puerto Rico. Full data for 2005 can be found at www.cdc.gov/asthma/nhis/05/table2-1.htm.

The association of *ADAM33* with asthma was first described by P. van Eerdewegh et al. (2002) *Nature* 418: 426-430 "Association of the ADAM33 gene with asthma and bronchial hyperresponsiveness." The latest research on the mechanism by which it works is described in R. G. del Mastro et al. (2007) *BMC Medical Genetics* 8:46 "Mechanistic role of a disease-associated genetic variant within the *ADAM33* asthma susceptibility gene."

For a review of inflammation and neonatal lung development see Shi, W., S. Bellusci, and D. Warburton (2007) *Chest* 132: 651-656 "Lung development and adult lung diseases."

The roles of T-cells in asthma are discussed in M. Larché (2007) *Chest* 132: 1007-1014 "Regulatory T cells in allergy and asthma," and Umetsu, D. T. and R. H. DeKruyff (2006) *Immunological Reviews* 212: 238-255 "The regulation of allergy and asthma," among many other places. An early review of the clinical applications of interleukin cytokines is P. J. Barnes (2001) *European Respiratory Journal* 34 (Suppl): 67-77 "Cytokine modulators as novel therapies for airway disease." Genetic associations with asthma are updated in Zhang, J., P. D. Pare, and A. J. Sandford (2008) *Respiratory Research* 9:4 "Recent advances in asthma genetics." The linkage with IRAK is described in L. Balaci et al. (2007) *American Journal of Human Genetics* 80: 1103-1114 "IRAK-M is involved in the pathogenesis of early-onset persistent asthma."

An intriguing study of the maternal genotype contribution of *HLA-G* and the possible role of microRNAs is by Z. Tan et al. (2007) *American Journal of Human Genetics* 81: 829-834 "Allele-specific targeting of microRNAs to HLA-G and risk of asthma." See also C. Ober (2005) *Immunology and Allergy Clinics of North America* 25: 669-679 "HLA-G: an asthma gene on chromosome 6p."

The genomewide association study is M. F. Moffatt et al. (2007) *Nature* 448: 470-473 "Genetic variants regulating *ORMDL3* expression contribute to the risk of childhood asthma."

inflamed bowels and crohn's disease

Genomewide association studies for Crohn's disease have been published by M. Parkes et al. (2007) *Nature Genetics* 39: 830-832 "Sequence variants in the autophagy gene IRGM and multiple other replicating loci contribute to Crohn's disease susceptibility," J. D. Rioux et al. (2007) *Nature Genetics* 39: 596-604 "Genomewide association study identifies new susceptibility loci for Crohn disease and implicates autophagy in disease pathogenesis," J. V. Raelson et al. (2007) *Proceedings of the National Academy of Science (USA)* 104: 14747-14752 "Genome-wide association study for Crohn's disease in the Quebec Founder Population identifies multiple validated disease loci," J. Hampe et al. (2007) *Nature Genetics* 39: 207-211 "A genome-wide association scan of nonsynonymous SNPs identifies a susceptibility variant for Crohn disease in ATG16L1," and A. Franke et al. (2007) *PLoS ONE* 2: e691 "Systematic association mapping identifies NELL1 as a novel IBD disease gene." For a review, see Xavier, R. J. and D. K. Podolsky (2007) *Nature* 448: 427-434 "Unraveling the pathogenesis of inflammatory bowel disease."

The cold chain hypothesis was proposed by Hugot, J. P., C. Alberti, D. Berrebi, E. Bingen, and J. P. Cézard (2003) *The Lancet* 362: 2012-2015 "Crohn's disease: the cold chain hypothesis."

A meta-analysis of the protective effect of hookworm infection against asthma is Leonardi-Bee, J., D. Pritchard, and J. Britton (2006) *American Journal of Respiratory and Critical Care Medicine* 174: 514-523 "Asthma and current intestinal parasite infection: systematic review and meta-analysis."

rheumatoid arthritis

A genomewide scan for rheumatoid arthritis is described by R. M. Plenge et al. (2007) *New England Journal of Medicine* 357: 1199-1209 "TRAF1-C5 as a risk locus for rheumatoid arthritis—a genomewide study."

The role of the HLA in mediating arthritis via immunity to citrulline is reviewed in Klareskog, L., J. Rönnelid, K. Lundberg, L. Padyukov, and L. Alfredsson (2008) *Annual Reviews of Immunology* 26: 651-675 "Immunity to citrullinated proteins in Rheumatoid Arthritis."

Nicotine's connection to arthritis is reviewed in Harel-Meir, M., Y. Sherer, and Y. Shoenfeld (2007) *Nature Clinical Practice. Rheumatology* 3: 707-715 "Tobacco smoking and autoimmune rheumatic diseases."

See Lie, B. A. and E. Thorsby (2005) *Current Opinion in Immunology* 17: 526-531 "Several genes in the extended human MHC contribute to predisposition to autoimmune diseases" for a brief discussion of the MHC in lupus. A more general review of autoimmunity is by A. M. Bowcock (2005) *Annual Review of Genomics and Human Genetics* 6: 93-122 "The genetics of psoriasis and autoimmunity."

imbalance of the immune system

The impact of selection on the promoter of the *IL4* gene is described by Rockman, M. V., M. W. Hahn, N. Soranzo, D. B. Goldstein, and G. A. Wray (2003) *Current Biology* 13: 2118-2123 "Positive selection on a human-specific transcription factor binding site regulating IL4 expression."

Chapter 5

AIDS and the world

Global estimates of AIDS prevalence can be found online at the following informative site: www.avert.org/worldstats.htm or at www.aidsmap.com. The latest epidemiology of AIDS is presented by J. Cohen (2007) *Science* 318: 1360-1361 "New estimates scale back scope of HIV/AIDS epidemic."

United States support policy and the ABCs of AIDS prevention are outlined by the USAID at www.usaid.gov/our_work/global_health/aids/News/abcfactsheet.html.

Al Gore's *The Assault on Reason* was published by Penguin books in 2007.

The South African AIDS portal is www.doh.gov.za/aids/index.html.

Journalist Jonny Steinberg has just completed a book about the epidemic: *Sizwe's Test: A Young Man's Journey through Africa's AIDS Epidemic* (Simon and Schuster, 2008).

For more on HAART, download the book at www.haart.com.

from HIV to AIDS

Severe Combined Immune Deficiency is a prime target for gene therapy in humans, as outlined for example in Cavazzana-Calvo, M., and A. Fischer (2007) *Journal of Clinical Investigation* 117: 1456-1465 "Gene therapy for severe combined immunodeficiency: are we there yet?"

AIDS-related cancer is reviewed in Arora, A., E. Chiao, and S. K. Tyring (2007) *Cancer Treatment Research* 133: 21-67 "AIDS malignancies." For more on Gardasil and HPV, see www.gardasil.com/.

A strong statement of how we know HIV causes AIDS is O'Brien, S. J. and J. J. Goedert (1996) *Current Opinion in Immunology* 8: 613-618 "HIV causes AIDS: Koch's postulates fulfilled." This is in response to the contrary opinion expressed by P. H. Duesberg (1988) *Science* 241: 514-517 "HIV is not the cause of AIDS" and elsewhere. Koch's postulates were initially published in German in 1891, but appear in T. M. Rivers (1937) *Journal of Bacteriology* 33: 1-12 "Viruses and Koch's postulates."

why HIV is so nasty

Drugs that inhibit the co-receptors that HIV uses to get into T-cells are reviewed in Biswas, P., G. Tambussi, and A. Lazzarin (2007) *Expert Opinion in Pharmacotherapy* 8: 923-933 "Access denied? The status of co-receptor inhibition to counter HIV entry," while the co-receptors themselves are described in Arenzana-Seisdedos, F. and M. Parmentier (2006) *Seminars in Immunology* 18: 387-403 "Genetics of resistance to HIV infection: Role of co-receptors and co-receptor ligands."

The evolution of resistance by HIV to even triple cocktails of drugs is described in N. Lohse et al. (2007) *Antiviral Therapy* 12: 909-917 "Genotypic drug resistance and long-term mortality in patients with triple-class antiretroviral drug failure."

Various AIDS medications are described at www.aidsmeds.com/list.shtml.

how to resist a virus with your genes

The ability of HIV to evolve resistance by switching from one co-receptor to another is established in Moncunill, G., M. Armand-Ugón, E. Pauls, B. Clotet, and J. A. Esté (2008) *AIDS* 22: 23-31 "HIV-1 escape to CCR5 coreceptor antagonism through selection of CXCR4-using variants in vitro."

The case for selection acting on CCR5 is challenged by P. C. Sabeti et al. (2005) *PLoS Biology* 3: e378. "The case for selection at CCR5-Delta32." For a discussion of selection and inclusive fitness, see P. Schliekelman (2007) *Evolution* 61: 1277-1288 "Kin selection and evolution of infectious disease resistance."

For more on Maraviroc, search for it at http://hivinsite.ucsf.edu.

The whole genome scan that turned up the associations between MHC and HIV traits is described in J. Fellay et al. (2007) *Science* 317: 944-947 "A whole-genome association study of major determinants for host control of HIV-1."

HIV imbalance

The feline and simian immunodeficiency viruses are discussed in S. VandeWoude and C. Apetrei (2006) *Clinical Microbiology Reviews* 19: 728-762 "Going wild: lessons from naturally occurring T-lymphotropic lentiviruses."

The argument that HIV was released as a result of the fight against polio was made in a book by Edward Hooper, *The River: A Journey Back to the Source of HIV and AIDS* (Penguin, 1999). The theory was refuted using evolutionary approaches by M. Worobey et al. (2004) *Nature* 428: 820 "Origin of AIDS: contaminated polio vaccine theory refuted."

The latest data on the origin of the two viral types can be found in B. F. Keele et al. (2006) *Science* 313: 523-526 "Chimpanzee reservoirs of pandemic and nonpandemic HIV-1," and Van Heuverswyn, F. and M. Peeters (2007) *Current Infectious Disease Reports* 9: 338-346 "The origins of HIV and implications for the global epidemic."

Chapter 6

creative depression

A useful Web site for sufferers of affective disorders is www.pendulum.org.

Many sites list celebrities who have suffered, including www.geocities.com/cover-bridge2k/artsci/famous_people_depression.html.

Much of the information on Winston Churchill, Harrison Ford, and others was gleaned from Wikipedia and linked sites.

The National Institute of Mental Health site is www.nimh.nih.gov/health/publications/depression/complete-publication.shtml, and the WHO's is www.who.int/mental_health/management/depression/definition/en.

an epidemic of mood swings

For a brief discussion of whether depression is epidemic, see D. Summerfield (2006) *Journal of the Royal Society of Medicine* 99: 161-162 "Depression: epidemic or pseudo-epidemic?"

Andrew Solomon's book is *The Noonday Demon: An Atlas of Depression* (Scribner, 2001).

bipolar and monopolar disorders

Wikipedia is as good a place as any to get some quick background: see http://en.wikipedia.org/wiki/Clinical_depression and http://en.wikipedia.org/wiki/Bipolar_disorder.

Major and bipolar depressive disorders are reviewed in Belmaker, R. H. and G. Agam (2008) *New England Journal of Medicine* 358: 55-68 "Major depressive disorder," and Miklowitz, D. J. and S. L. Johnson (2006) *Annual Review of Clinical Psychology* 2: 199-235 "The psychopathology and treatment of bipolar disorder," respectively.

The genetic component of depression is established by McGuffin, P., F. Rijsdijk, M. Andrew, P. Sham, R. Katz, and A. Cardno (2003) *Archives of General Psychiatry* 60: 497-502 "The heritability of bipolar affective disorder and the genetic relationship to unipolar disorder."

A. Halfin (2007) *American Journal of Managed Care* 13: (4 Suppl) S92-S97 "Depression: the benefits of early and appropriate treatment" covers the medical and financial costs of failure to treat depression when it first appears.

the pharmacology of despair

The role of serotonin in depression is reviewed in Ressler, K. J. and C. B. Nemeroff (2000) *Depression and Anxiety* 12 (Suppl 1): 2-19 "Role of serotonergic and noradrenergic systems in the pathophysiology of depression and anxiety disorders."

That of cortisol can be found in Thomson, F. and M. Craighead (2007) *Neurochemical Research* 33: 691-707) "Innovative Approaches for the Treatment of Depression: Targeting the HPA Axis."

Antidepressants are considered in J. J. Mann (2005) *New England Journal of Medicine* 353: 1819-1834 "The medical management of depression," and H. J. Gijsman (2004) *American Journal of Psychiatry* 161: 1537-1547 "Antidepressants for bipolar depression: a systematic review of randomized, controlled trials."

misbehaving serotonin

The two studies describing interactions between genes and culture are Caspi, A. et al. (2002) *Science* 297: 851-854 "Role of genotype in the cycle of violence in maltreated children," and Caspi, A. et al. (2003) *Science* 301: 386-389 "Influence of life stress on depression: Moderation by a polymorphism in the 5-HTT gene."

For more on genes and suicide, see Bondy, B., A. Buettner, and P. Zill (2006) *Molecular Psychiatry* 11: 336-351 "Genetics of suicide."

The hypothesis that depression may be due to serotonin resistance is articulated by Smolin, B., E. Klein, Y. Levy, and D. Ben-Shachar (2007) *International Journal of Neuropsychopharmacology* 10: 839-850 "Major depression as a disorder of serotonin resistance: inference from diabetes mellitus type II."

faint genetic signals

The genetics of depression is reviewed by D. F. Levinson (2006) *Biological Psychiatry* 60: 84-92 "The genetics of depression: a review," and of bipolar disorder by T. Kato (2007) *Psychiatry and Clinical Neurosciences* 61: 3-19 "Molecular genetics of bipolar disorder and depression."

A linkage scan for bipolar disorder is described in M. B. McQueen et al. (2005) *American Journal of Human Genetics* 77: 582-595 "Combined analysis from eleven linkage studies of bipolar disorder provides strong evidence of susceptibility loci on chromosomes 6q and 8q." BD is one of the diseases that failed to show any associations in Wellcome Trust Case Control Consortium (2007) *Nature* 447: 661-678 "Genome-wide association study of 14,000 cases of seven common diseases and 3,000 shared controls."

The connection to variation in the serotonin transporter gene is reviewed in Uher, R. and P. McGuffin (2007) *Molecular Psychiatry* 13: 131-146 "The moderation by the serotonin transporter gene of environmental adversity in the aetiology of mental illness: review and methodological analysis," and in H. A. Mansour et al. (2005) *Annals of Medicine* 37: 590-602 "Serotonin gene polymorphisms and bipolar I disorder: focus on the serotonin transporter."

schizophrenic spectrum and other mental disturbances

Iceland's deCODE Genetics was reported by Michael Specter in *The New Yorker* (January 18, 1999, 40-51) "Decoding Iceland: the next big medical breakthroughs may result from one scientist's battle to map the Viking gene pool."

One of the deCODE studies of schizophrenia is Stefansson, H., V. Steinthorsdottir, T. E. Thorgeirsson, J. Gulcher, and K. Stefansson (2004) *Annals of Medicine* 36: 62-71 "Neuregulin 1 and schizophrenia." The same group, and another consortium, discuss the contribution of CNV to schizophrenia in two papers in *Nature* released online on July 30, 2008, by H. Stefansson et al., "Large recurrent microdeletions associated with schizophrenia," and by The International Schizophrenia Consortium, "Rare chromosomal deletions and duplications increase risk of schizophrenia."

Cannon, T. D. and M. C. Keller (2006) *Annual Review of Clinical Psychology* 2: 267-290 "Endophenotypes in the genetic analyses of mental disorders" describes how it may be more fruitful to study attributes related to schizophrenia than the disease itself.

For an update on mental retardation see Debacker, K. and R. F. Kooy (2007) *Human Molecular Genetics* 2007 16 (Spec 2): R150-R158 "Fragile sites and human disease." An intriguing link between common deletions and autism was reported by J. Sebat et al. (2007) *Science* 316: 445-449 "Strong association of de novo copy number mutations with autism."

the genetic tightrope of the mind

The books referenced in this section are Tom Wolfe's *I am Charlotte Simmons: A Novel* (Farrar, Straus and Giroux, 2004) and Martha Stout's *The Paranoia Switch: How Terror Rewires Our Brains and Reshapes Our Behavior—and How We Can Reclaim Our Courage* (Farrar, Straus and Giroux, 2007).

a kindling theory for the modern world

The kindling theory is evaluated in Kendler, K. S., L. M. Thornton, and C. O. Gardner (2000) *American Journal of Psychiatry* 157: 1243-1251 "Stressful life events and previous episodes in the etiology of major depression in women: an evaluation of the 'kindling' hypothesis." It traces back to Emil Kraepelin: *Manic-Depressive Insanity and Paranoia (Foundations of Modern Psychiatry)* available in English from Thoemmes Continuum (2002), and was articulated by Segal, Z. V., J. M. Williams, J. D. Teasdale, and M. Gemar (1996) *Psychological Medicine* 26: 371-380 "A cognitive science perspective on kindling and episode sensitization in recurrent affective disorder."

For an account of flies on cocaine, see Andretic, R., S. Chaney, and J. Hirsh (1999) *Science* 285: 1066-1068 "Requirement of circadian genes for cocaine sensitization in Drosophila." Kindling in epilepsy and bipolar disorder is discussed in Bertram, E. (2007) *Epilepsia* 48 (Suppl 2): 65-74 "The relevance of kindling for human epilepsy," and in Amann, B. and H. Grunze (2005) *Epilepsia* 46 (Suppl 4): 26-30 "Neurochemical underpinnings in bipolar disorder and epilepsy."

For overviews of the addiction, see Carlton K. Erickson *The Science of Addiction: From Neurobiology to Treatment* (W.W. Norton, 2007), or the companion book to an HBO television series on the topic, John Hoffman and Susan Froemke's *Addiction, Why Can't They Just Stop? New Knowledge, New Treatments, New Hope.*

Chapter 7

slow walk to dementia

For a satirical view opinion about Ronald Reagan and Alzheimer's, see www.theonion.com/content/node/27646. An actual Congressional effort to aid research in the name of the former President is an Act introduced in 2004: http://olpa.od.nih.gov/legislation/108/pendinglegislation/reagonalzheimer.asp.

An analysis of healthcare costs is by Alemayehu, B. and K. E. Warner (2004) *Health Services Research* 39: 627-642 "The lifetime distribution of healthcare costs."

Elan's AN1792 vaccine is described in Schenk, D. B., P. Seubert, M. Grundman, and R. Black (2005) *Neurodegenerative Disease* 2: 255-260 "A beta immunotherapy: Lessons learned for potential treatment of Alzheimer's disease."

alzheimer's on the march

See Goedert, M. and B. Ghetti (2007) *Brain Pathology* 17: 57-62 "Alois Alzheimer: his life and times" for more about the man who first described AD.

Global population statistics and age pyramids for every country can be downloaded at www.census.gov/ipc/www/idb/.

An opinion piece on the epidemiology of AD is C. Brayne (2007) *Nature Reviews* 8: 233-239 "The elephant in the room—healthy brains in later life, epidemiology and public health."

tangles and plaques

For recent discussions of the roles of Aβ and Tao in AD see Cappai, R. and K. J. Barnham (2008) *Neurochemical Research* 33: 526-532 "Delineating the mechanism of Alzheimer's disease Aβ peptide neurotoxicity," and Ballatore, C., V. M. Lee, and J. Q. Trojanowski (2007) *Nature Reviews* 8: 663-672 "Tau-mediated neurodegeneration in Alzheimer's disease and related disorders."

The correct identification of Alzheimer's patients before death is important for the development of new drugs in clinical trials; new methods are described in Cummings, J. L., R. Doody, C. Clark (2007) *Neurology* 69: 1622-1634 "Disease-modifying therapies for Alzheimer disease: challenges to early intervention." Efforts to track the course of dementia are described in R. Lambon et al. (2003) *Brain* 126: 2350-2362 "Homogeneity and heterogeneity in mild cognitive impairment and Alzheimer's disease: a cross-sectional and longitudinal study of 55 cases."

early onset FAD

The diversity of Presenilin mutations is reviewed in M. Menéndez (2004) *Journal of Alzheimer's Disease* 6: 475-482 "Pathological and clinical heterogeneity of presenilin 1 gene mutations." The spectrum of APP mutations is described in A. Kowalska (2003) *Folia Neuropathology* 41: 35-40 "Amyloid precursor protein gene mutations responsible for early-onset autosomal dominant Alzheimer's disease." An old review of the genetics of FAD is P. H. St. George-Hyslop (2000) *Biological Psychiatry* 47: 183-199 "Molecular genetics of Alzheimer's disease."

For the latest on AD and Down syndrome, see Visootsak, J. and S. Sherman (2007) *Current Psychiatry Reports* 9: 135-140 "Neuropsychiatric and behavioral aspects of trisomy 21."

late onset LOAD

The original report of an association between ApoE and AD was W. J. Strittmatter et al. (1993) *Proceedings of the National Academy of Science USA* 90: 1977-1981 "Apolipoprotein E: high-avidity binding to beta-amyloid and increased frequency of type 4 allele in late-onset familial Alzheimer disease." Singh, P. P., M. Singh, and S. S. Mastana (2006) *Annals of Human Biology* 33: 279-308 "APOE distribution in world populations with new data from India and the UK" describe how LOAD genetic susceptibility due to ApoE is distributed around the globe.

The disconcordance between African and African American rates of AD is reported in Hendrie, H. C., J. Murrell, S. Gao, F. W. Unverzagt, A. Ogunniyi, and K. S. Hall (2006) *Alzheimer Disease and Associated Disorders* 20(Suppl 2): S42-S46 "International studies in dementia with particular emphasis on populations of African origin."

A genomewide association study for LOAD has just been published: H. Li et al. (2008) *Archives of Neurology* 65: 45-53 "Candidate single-nucleotide polymorphisms from a genomewide association study of Alzheimer disease." It has not yet led to any more really convincing leads.

just growing old

A list of animal life spans is at www.wonderquest.com/LifeSpan-MaxMin.htm.

A review of aging in model organisms is A. Antebi (2007) *PLoS Genetics* 3: e129 "Genetics of aging in *Caenorhabditis elegans*." The role of FOXO in mediating stress is reviewed in van der Horst, A. and B. M. Burgering (2007) *Nature Reviews Molecular Cell Biology* 8: 440-450 "Stressing the role of FoxO proteins in lifespan and disease."

A good online introduction to evolutionary theories of aging is by L. A. Gavrilov and N. S. Gavrilova (2002) *The Scientific World Journal* 2: 339-356 "Evolutionary theories of aging and longevity," also at http://longevity-science.org/Evolution.htm.

The mutation accumulation theory is generally attributed to Nobel Laureate Peter Medawar, who introduced it in *An Unsolved Problem of Biology* (H. K. Lewis, London, 1952). The antagonistic pleiotropy theory was formalized by Williams, G. C. (1957) *Evolution* 11: 398-411 " Pleiotropy, natural selection and the evolution of senescence."

http://en.wikipedia.org/wiki/Life_expectancy is an interesting site, though it is not clear where all the data and estimates come from.

Chapter 8

height and weight

The first phase genome scan for height was published by M. Weedon et al. (2007) *Nature Genetics* 39: 1245-1250 "A common variant of *HMGA2* is associated with adult and childhood height in the general population." A follow-up has been submitted, while a second group presented S. Sanna et al. (2008) *Nature Genetics* 40: 198-203 "Common variants in the GDF5-UQCC region are associated with variation in human height."

Canine size is described in N. B. Sutter et al. (2007) *Science* 316: 112-115 "A single *IGF1* allele is a major determinant of small size in dogs." Spady, T. C. and E. A. Ostrander (2008) *American Journal of Human Genetics* 82: 10-18 "Canine behavioral genetics: pointing out the phenotypes and herding up the genes" review the genetics of dog behavior.

Several references to *FTO* and obesity were given in the notes to Chapter 3, "Not so Thrifty Diabetes Genes." but interestingly, one that just came out suggests a difference in Chinese populations: H. Li et al. (2008) *Diabetes* 57: 264-268 "Variants in the fat mass- and obesity-associated (FTO) gene are not associated with obesity in a Chinese Han population."

pigmentation

For a thoughtful essay on human pigmentation, see E. J. Parra (2007) *American Journal of Physical Anthropology* 45 (Suppl): 85-105 "Human pigmentation variation: evolution, genetic basis, and implications for public health." Selection on pigmentation genes is described in Lao, O., J. de Gruijter, K. van Duijn, A. Navarro, and M. Kayser (2007) *Annals of Human Genetics* 71(Pt 3): 354-369 "Signatures of positive selection in genes associated with human skin pigmentation as revealed from analyses of single nucleotide polymorphisms." See also a meeting review of the American Association of Physical Anthroplogists by Ann Gibbons (2007) *Science* 316: 364 "European skin turned pale only recently, gene suggests" *Science* 316: 364.

The genomewide association for eye and hair color is P. Sulem et al. (2007) *Nature Genetics* 39: 1443-1452 "Genetic determinants of hair, eye and skin pigmentation in Europeans." One for skin color in Asians is R. P. Stokowski et al. (2007) *American*

Journal of Human Genetics 81: 1119-1132 "A genomewide association study of skin pigmentation in a South Asian population."

How a gene identified in fish was shown to impact human skin color is described in Lamason, R.L. et al. (2005) *Science* 310: 1782-1786 "*SLC24A5*, a putative cation exchanger, affects pigmentation in zebrafish and humans." The reference to Neanderthal skin color is C. Lalueza-Fox et al. (2007) *Science* 318: 1453-1455 "A melanocortin 1 receptor allele suggests varying pigmentation among Neanderthals."

Consumers are warned about the relationship between sunlight, vitamin D, and osteoporosis at www.osteoporosis.org.au/files/internal/CalciumVitD_consumer.pdf.

the God gene

The Web sites for personal genomics are https://www.23andme.com, www.decodeme.com, and www.personalgenomes.org.

J. Craig Venter's book is *A Life Decoded: My Genome: My Life* (Viking Adult, 2007).

The paper describing his own genome sequence is S. Levy *et al.* (2007) *PLoS Biology* 5: e254 "The diploid genome sequence of an individual human."

James Watson's sequencing project can be accessed at www.454.com/watson/. For the *New York Times* viewpoint, see www.nytimes.com/2007/05/31/science/31cnd-gene.html

Dean Hamer's two books are *The God Gene: How Faith Is Hardwired into Our Genes* (Anchor, 2005) and *The Science of Desire: The Gay Gene and the Biology of Behavior* (Touchstone, 1995).

The heritability of spirituality measure is explored in Kirk, K. M., L. J. Eaves, and N. G. Martin (1999) *Twin Research* 2: 81-87 "Self-transcendence as a measure of spirituality in a sample of older Australian twins." To my knowledge, the association with *VMAT2* has never been published in a peer-reviewed journal. Hamer's group reports their latest on the genetics of sexual orientation in B. S. Mustanski et al. (2005) *Human Genetics* 116: 272-278 "A genomewide scan of male sexual orientation."

a few words about IQ

The paper that showed selection on Microcephalin is P. D. Evans et al. (2005) *Science* 309: 1717-1720 "*Microcephalin*, a gene regulating brain size, continues to evolve adaptively in humans," while the linkage to IQ is refuted by N. Bekel-Bobrov et al. (2007) *Human Molecular Genetics* 16: 600-608 "The ongoing adaptive evolution of *ASPM* and *Microcephalin* is not explained by increased intelligence."

A controversial book that touches on genetics, race, and IQ is Herrnstein, R. J. and C. Murray *The Bell Curve: Intelligence and Class Structure in American Life* (Free Press, 1994). For a series of articles on this matter, see *American Psychologist* Volume 60, Issue 1 (Jan 2005), or for an example of a blog, see http://shrinkwrapped.blogs.com/blog/2007/11/iq-genetics-and.html.

The link between breast-feeding, *FAS2*, and IQ is reported in A. Caspi et al. (2007) *Proceedings of the National Academy of Science USA* 104: 18860-18865 "Moderation of breastfeeding effects on the IQ by genetic variation in fatty acid metabolism."

on being human

Several studies detect selection on parts of the human genome, as reviewed in Nielsen, R., I. Hellmann, M. Hubisz, C. Bustamante, and A. G. Clark (2007) *Nature Reviews Genetics* 8: 857-868 "Recent and ongoing selection in the human genome" and Sabetti, P. C. et al. (2006) *Science* 312: 1614-1620 "Positive natural selection in the human lineage." See also Clark, A. G. et al. (2003) *Science* 302: 1960-1963 "Inferring nonneutral evolution from human-chimp-mouse orthologous gene trios," C. D. Bustamante et al. *Nature* 437: 1153-1157 "Natural selection on protein coding genes in the human genome," and Haygood, R., O. Fedrigo, B. Hanson, K. Yokoyama, and G. A. Wray (2007) *Nature Genetics* 39:1140-1144 "Promoter regions of many neural- and nutrition-related genes have experienced positive selection during human evolution."

The first of several studies of gene expression in human and chimpanzee brains was W. Enard et al. (2002) *Science* 296: 340-343 "Intra- and interspecific variation in primate gene expression patterns." These are reviewed in Khaitovich, P., W. Enard, M. Lachmann, and S. Pääbo (2006) *Nature Reviews Genetics* 7: 693-702 "Evolution of primate gene expression."

the adolescent genome revisited

A scientific synopsis of the argument in this book is in press: Gibson, G. (February 2009) *Nature Reviews Genetics* "Decanalization and the Origin of Complex Disease."

About the author

Greg Gibson is Professor of Genetics at North Carolina State University in Raleigh, and of Integrative Biology at the University of Queensland, Australia. He is a leader in the new field of genomics, studying how interactions between genes and the environment affect human health and organismal evolution. He holds a Ph.D. from the University of Basel, Switzerland, and did postdoctoral work at Stanford University. He is on the editorial boards of *PLoS Genetics*, *Current Biology*, *Genetics*, and other leading journals, and with Spencer V. Muse, coauthored *A Primer of Genome Science*, one of the field's leading textbooks, now in its third edition.

Index